Unfolding Crisis in Assam's Tea Plantations

Transition in Northeastern India

Series Editor: **Sumi Krishna**, Independent scholar, Bangalore

The uniquely diverse landscapes, societies and cultures of northeastern India, forged through complex bio-geographic and socio-political forces, are now facing rapid transition. Yet, popular and academic perceptions tend to be limited primarily to the various conflicts in the region. This series, therefore, aims to broaden the focus to the processes and practices that have shaped, and are shaping, the peoples' identities, outlook, institutions and economy. Eschewing the homogenising term 'North East', which was imposed on the region in a particular political context half a century ago, the series title refers to the 'northeastern' region to more accurately reflect its heterogeneity and the varied issues confronting its diverse peoples. The series will encompass a broad rubric of themes related to culture, social relations, human and economic development, the environment, technology, governance and juridical systems.

Seeking to explore how the 'mainstream' and the 'margins' impact each other, the series will foreground both historical and contemporary research on the northeastern region including the Eastern Himalaya, the adjoining hills and valleys, the states of Arunachal Pradesh, Assam, Manipur, Meghalaya, Mizoram, Nagaland, Sikkim and Tripura. It will publish original, reflective studies that draw upon different disciplines and approaches, and combine empirical and theoretical insights. The monographs and the occasional edited volume are intended to make scholarship accessible for a wide spectrum of general readers and to help deepen the understanding of academics, policy-makers and practitioners.

Also in this Series

Education and Society in a Changing Mizoram: The Practice of Pedagogy
Lakshmi Bhatia
978-0-415-58920-8

Becoming a Borderland: Space and Identity in Colonial Northeastern India
Sanghamitra Misra
978-0-415-61253-1

Unfolding Crisis in Assam's Tea Plantations

Employment and Occupational Mobility

Deepak K. Mishra
Vandana Upadhyay
Atul Sarma

Routledge
Taylor & Francis Group

LONDON NEW YORK NEW DELHI

First published 2012 in India
by Routledge
912 Tolstoy House, 15–17 Tolstoy Marg, Connaught Place, New Delhi 110 001

Simultaneously published in the UK
by Routledge
2 Park Square, Milton Park, Abingdon, Oxfordshire OX14 4RN

First issued in paperback 2015

Routledge is an imprint of the Taylor & Francis Group, an informa business

Typeset by
Star Compugraphics Private Limited
5, CSC, Near City Apartments
Vasundhara Enclave
Delhi 110 096

British Library Cataloguing-in-Publication Data
A catalogue record of this book is available from the British Library

ISBN 13: 978-1-138-66254-4 (pbk)
ISBN 13: 978-0-415-52308-0 (hbk)

Contents

List of Tables

List of Figures

Acknowledgements

This book is based on the findings of a research project entitled 'Occupational Mobility of Plantation Sector Labours in Assam: Determinants and Implications' carried out at the Department of Economics, Rajiv Gandhi University (formerly Arunachal University), Itanagar, with financial support from the National Tea Research Foundation (NTRF), Kolkata. We would like to express our profound gratitude to Dr Barundeb Banerjee, advisor, NTRF and Basudeb Banerjee, chairman, Tea Board, Kolkata, for all the help and support extended to us during our research. We are grateful to Mr Naba Kumar Das, former Chairman, Tea Board and currently Chief Secretary to the Government of Assam for his active interest in this study. This work was greatly facilitated by the informed counsel and support given to us by the officials of the NTRF as also those of the Tea Board, Kolkata. We thank A.R. Sarkar, research officer, NTRF, for co-ordinating the work and extending to us all possible support. Special thanks also to M. Paramanantham, statistician, and Nasrullah Syed of Tea Board, Kolkata, for providing us with all the statistics and secondary data for our work.

Sumi Krishna, the series editor of this volume, was a great support and her involvement in this book extends beyond the normal responsibilities of a series editor. Not only did she take an active interest in this study, she also provided valuable feedback which makes this book more accessible to a wider range of readers. We have benefited from the comments and observations of participants at the Annual Conference of the British Association of South Asian Studies at the University of Warwick, where a part of the study was presented. We are deeply indebted to Prof Barbara Harriss-White, University of Oxford, for her insightful comments and suggestions.

We are especially indebted to our research fellow Dr Sewali Kurmi for her fieldwork and constant support at various stages. It would not have been possible for us to complete the work without her total involvement. Special thanks are also due to Dr Jawan Singh Rawat for his excellent research support and for compiling the data.

We gratefully acknowledge his assistance. We also acknowledge the help of Deepak Kumar, research scholar, Centre for the Study of Regional Development, Jawaharlal Nehru University for his help in data analysis. Thanks are also due to all the field assistants, especially to Kamakhaya Prasad and Dr Robin Khataniar, who have helped us in data collection. We are also grateful to Dr R. Thapa, Meghali Baruah and Dr Archana Upadhyay for their generous help during the field survey and also for the valuable discussions that we had with them regarding the conditions of the tea garden labourers. Dr Kalyan Das, OKD Institute of Social Change, Guwahati, who himself has worked on various dimensions of the tea sector in Assam and also in Sri Lanka, was generous enough to share his unpublished and published research on the topic and we are thankful to him.

The help and support of Debasish Chakraborty, Secretary, ABITA, Tezpur and the managers of the tea gardens of Panitola Tea Estate, Dibrugarh, Harmotty Tea Estate, Lakhimpur and Nahorhabi Tea Estate, Sivasagar in Assam is gratefully acknowledged. We would like to thank the editorial team at Routledge, for their valuable suggestions on earlier drafts of the book. Last but not the least, we express our gratitude to the workers and villagers, who not only spared their valuable time to answer our queries, but also extended warm hospitality to the research team. Needless to add, the sole responsibility for mistakes and shortcomings lies with the authors.

1

Introduction

The tea industry occupies a significant position in the history of the development of the Indian economy in terms of its contribution in exports, income and employment. The tea sector is said to be undergoing a crisis since the early 1990s. This crisis in the tea industry is manifested through stagnation in production, decline in exports and closure of tea gardens. Although its linkage to the opening up of the economy has been widely commented upon, at the root of the crisis lies low productivity in the tea sector. There is a growing body of opinion that argues that high labour costs — a result of state regulations to ensure some security to workers — are the root cause of the failure of the Indian tea industry to be competitive in both the global and domestic market. As tea estates are being closed down and labour is being retrenched to cut down costs, tea garden labourers are facing a crisis of livelihood. For a number of reasons, it is difficult for them to move out of the gardens and find alternative sources of livelihoods. Their story is among many such contradictions that remain at the margins of the narratives of a globalising and 'shining' India.

India's recent growth performance is widely credited to the bold steps that were taken to deregulate and liberalise the economy with the reforms initiated in 1991. While the overall growth of the economy has fuelled considerable optimism within and outside the country, India's agriculture sector has failed to grow at an appreciable rate. In fact, productivity growth in agriculture has decelerated since the 1990s, leading to an agrarian crisis in many of the poorer regions of India. The informal sector that provides employment to nearly 90 per cent of the workers has also been showing signs of continuing distress in this period of rapid economic growth. Nearly 70 per cent of those who depend on the informal sector have been estimated to be poor. There are indications that as the Indian economy integrates further into the global circuits of production, exchange and accumulation, the burden of adjustments is being disproportionately shared by small and marginal farmers, agricultural labourers and workers in the urban informal sector. This study, though focused on a much narrower question, attempts to understand the work conditions in Assam's tea plantations that share the characteristics of both agriculture and industry. Although the tea sector is officially part of the organised sector, informalisation and casualisation of labour have been among the key dimensions of the recent changes in this sector.

Historically, the plantation economies in general and the tea sector in particular have been associated with bondage and indenture labour systems, implying varying degrees of unfreedom for the labourers. Notwithstanding the state-initiated efforts to safeguard the constitutionally guaranteed minimal rights to the workers, reorganisation of the sector in response to changing market conditions and near-complete unionisation of the workers in this sector, there seems to be limited mobility of tea garden workers (or ex-tea garden workers) or their descendants in terms of diversification of sources of earnings and employment. There seems to be a considerable degree of 'crowding in' of tea garden workers and their families in the tea sector itself. In the context of rapid and increasing mobility of workers across sectors and spaces, a process that has gained further momentum with increasing global integration of economies, the apparent intergenerational immobility of tea garden labourers is an important and, to some extent, intriguing question that deserves detailed research attention.

This study is an attempt to investigate some important dimensions of labour market characteristics in the tea sector. Although it is a study of the tea sector during a period of crisis, it is not an investigation into the possible causes of the crisis. The focus is on the labour market and the implications of the ongoing crisis and restructuring of the sector for the tea garden labourers and their families. While investigating the changes in the gardens through the prism of labour relations and employment conditions in the tea gardens, the study raises several questions that have significance, we believe, beyond the spatio-temporal context. It reveals the way forces of globalisation and neo-liberal reforms have been reshaping the world of work of the labourers at the margins, at the same time we have not followed a standard comparison between pre- and post-reform periods. The issues that we have investigated here — labour relations, occupational mobility, livelihoods diversification and intergenerational changes — go beyond the changes that have taken place in the recent past. Without explicitly going into a historical analysis of the way labour relations have evolved in the gardens, we have attempted to understand and contextualise the changes in conditions of work in relation to the specific forms of labour organisation in gardens, which of course have been shaped, to an extent, by the historical specificities of the tea sector in Assam. In particular, we focus on the following questions. What is the extent of intergenerational mobility and occupational diversification among the tea garden labourers? What is the nature, direction and pattern of occupational diversification, if any? Is it distress-induced or a response to newer and better opportunities? If there is little diversification and most of the workers and their descendants are 'crowding in' in the tea sector itself, is the immobility at the tea-sector level or are the workers tied to specific tea gardens and employers? Is occupational immobility associated with any mobility in the income ladder or not? What are the underlying causes of the relative immobility or diversification of occupations at the household level? What are the implications of all these for the households' survival and well-being? And, finally, what are the policy implications of these developments?

The Tea Industry in India

Tea is India's oldest industry in the organised manufacturing sector. It has continued to maintain its position as the single largest employer in this sector. One out of seven workers in organised manufacturing is a tea plantation worker. India is the largest producer of tea in the world, producing around 790 million kg annually (Tea Board 2004). Around 30 per cent of global tea production is produced in India. India is also the world's largest consumer of tea. Among the four largest tea producers in the world, viz. India, Bangladesh, Sri Lanka and Kenya, India exports the lowest amount of tea.

The tea industry in India began with the founding of the Assam Company in 1839, although the potential for growing tea was discovered earlier, in 1824, by Major Robert Bruce when he came across indigenous tea bushes in Assam (ITA 1933 cited in Bhowmik 1981).

Since then, the industry has seen continuous expansion and con-solidation during the colonial period. At present, the four major tea producing states in the country are Assam, West Bengal, Tamil Nadu and Kerala. Today, Assam produces around 52 per cent of the total tea produced in India and employs around 56 per cent of the labour force working in this sector (Tea Board 2004). Among the different districts of Assam, Dibrugarh alone accounts for 34 per cent of the total tea produced in the state, while around 26 per cent of the total production comes from Sivasagar district.

The tea industry in India is said to be facing a severe crisis, particularly after the disintegration of the Soviet Union, the largest importer of Indian tea. However, over the past decade, domestic consumption of tea has increased at a faster rate than production — at a steady rate of around 15 to 20 million kg annually. The steady increase in domestic demand and the inability of the tea sector to enhance production has resulted in a decline in tea exports. The decision of the government to allow cheaper tea imports from Bangladesh and Sri Lanka, according to tea producers, has only deepened the crisis (Bhowmik 2002).

While low labour productivity is frequently cited as the main reason behind the crisis faced by the sector, other variables such as inability to expand the area under cultivation, ageing of the tea bushes,[1] inadequate replanting of bushes, inadequate investments in plant modernisation and labour welfare measures,[2] and traditional, cost-ineffective management practices have also contributed towards the near-stagnation of production (Sivaram 2000; Bhowmik 2002). The present study attempts to investigate the trends in production, productivity and employment in the tea sector, particularly in Assam. As a backdrop to the analysis we describe the broader context in which these changes have occurred.

---

[1]Normally the tea bush achieves its prime in terms of output of green leaves from the age of five to 30 years. In 2002, 35 per cent of tea bushes in India were over 50 years old, considered the uneconomic phase of a tea bush. In Assam, 29 per cent of tea bushes were over 50 years old (see Table 2.36 in Chapter 2; also see Bhowmik 2002).

[2] For a discussion on the evidences showing the relationship between investment in the health and welfare of workers on labour productivity in the tea sector see Sivaram (2000).

Labour and Globalisation

Globalisation, as a process of economic and non-economic restructuring, involves increasing economic integration among national economies.[3] Several dimensions of globalisation have been noted in the literature[4] (Beck 2000; Helleiner 2000). Globalisation has been distinguished in terms of its three different manifestations:

> [F]irst, the multiplication and intensification of economic, political, social, and cultural linkages among people, organizations and countries at the world level; second, the tendency towards the universal application of economic, institutional, legal, political, and cultural practices; and third, the emergence of significant spillovers from the behaviour of individuals and societies to the rest of the world. (von Braun and Diaz-Bonilla 2008: 4–5).

In some of these formulations, globalisation has been conceptualised as a process of fundamental restructuring of the global economy in

---

[3] We do not intend to provide a summary of the rich debate on the meanings of the term globalisation and its relevance as a concept (see Held and McGrew 2003; Scholte 1997; Hirst 1997; Hirst and Thompson 1996). Defining globalisation has itself become part of the contestations around the economic and social processes associated with it. The International Monetary Fund (IMF) website, for example, while stressing that '[t]here is substantial evidence, from countries of different sizes and different regions, that as countries "globalize" their citizens benefit, in the form of access to a wider variety of goods and services, lower prices, more and better-paying jobs, improved health, and higher overall living standards', defines globalisation in the following words: 'Economic "globalization" is a historical process, the result of human innovation and technological progress. It refers to the increasing integration of economies around the world, particularly through the movement of goods, services, and capital across borders. The term sometimes also refers to the movement of people (labor) and knowledge (technology) across international borders. There are also broader cultural, political, and environmental dimensions of globalization' (IMF 2008).

[4] Globalisation literature has focused on several dimensions of it. While broader definitions include social, political and cultural dimensions along with the economic aspects, the narrower definitions focus only on the economic aspects. Held et al. (1999: 1) have argued: 'globalization is in danger of becoming, if it has not already become, the cliché of our times: the big idea which encompasses everything from global financial markets to the Internet but which delivers little substantive insight into the contemporary human condition.'

historically unprecedented, irreversible ways (Skonieczny 2010). Many authors have, however, questioned this conceptualisation of globalisation as a historically unique process of change. They have attempted to show that in its essence, globalisation represents a relatively advanced phase of capitalism, which has strong historical connections (Hirst 1997; Sweezy 1997). Of course, many new developments of the global economy have impacted the global political economy in diverse ways, but these changes have been analysed with reference to the historical context under which capitalism has developed and has consolidated its position across the world. By its very nature, capitalism is an expanding system. It has transformed non-capitalist forms of production and distribution across the world. In its attempt to bring newer spaces under the overarching logic of capitalist production it has been reaching out to new areas as well as newer aspects of life and livelihoods.

> While acknowledging that most pre-capitalist economic forma-
> tions exhibited elements of commodity exchange to one degree or
> another, he argued that capitalism differentiates itself as a fully
> operational commodity system by virtue of two interdependent
> processes having reached a critical stage of development: a 'stretching'
> of commodity relations in both a territorial and quantitative sense to
> the point where production for a regional or national market displaces
> subsistence production as the primary form; and a 'deepening' of
> commodity relations in the sense that these now encompass not
> only goods and services but also labour power itself (Lysandrou
> 2005: 774).

International flow of goods, services and capital has been one of the most remarkable aspects of capitalist growth in the 19th century. Some authors, in this context, have attempted to argue that it was the post-war decades which saw the rise of protectionist trade regimes and Keynesian macroeconomic management of national economies which were an exception. Otherwise, international trade and trans-border movement of capital have always been an integral part of the capitalist system. In the renewed emphasis on trade and trans-border economic flows since the 1990s they see a necessary connection between the pre-second world economic order and the post-Soviet world economic order.

[I]t is worth recognizing that the recent splurge in globalization is part of an ongoing process with a long history. To begin with, capitalism was born in the process of creating a world market, and the long waves of growth in the core capitalist countries were associated with centuries-long spread by conquest and economic penetration. In the past as in the present, competitive pressures, the incessant need for capital to keep on accumulating, and the advantages of controlling raw material sources have spurred business enterprise to reach beyond its national borders. The tempo and nature of expansion has of course varied over time... (Magdoff 2002: 278).

Notwithstanding the historical connections with the past, this phase of capitalist world order had its own specific features. The scale of international flow of commodities, services and capital, the phenomenal improvement in transport and communication technologies, the rise of telecommunications and information and communications technologies (ICTs), and the significant rise of financial capital are all features of contemporary global capitalism that bring to it new dimensions. The emergence of the New World Order, after the demise of the Soviet Union, has led to changes in the political and economic management in such a way that capitalist market economies, combined with liberal democracy, are celebrated as the only viable and desirable form of political and economic organisation. The last decades of the 20th century were marked by an unprecedented ascendance of neo-liberal ideas. There were attacks on labour rights of various kinds and all the variants of social democracy that had emerged across advanced capitalist countries initiated sweeping reforms. Summarising the implications of these developments in the developed world, Andrew Glyn (2006: 49) notes:

Complementing the turn to financial austerity, the degree of government intervention in the dominant market sector of the economy has been drastically cut through privatization and latterly reductions in a wide range of product market regulations. Finally there has been a forceful campaign from the international organizations in favour of freeing up market forces in the labour market by cutting unemployment benefits, minimum wages and employment protection, the hard-won gains of the 1960s and 1970s.

Under the competitive pressure of globally mobile capital, and under the influence of global financial institutions like World Bank and the International Monetary Fund (IMF), third world countries started adopting measures that catered to the demands of capital. In order to emerge as, or remain, an attractive destination for capital, domestic deregulation and trade liberalisation policies were adopted by many developing countries across the board. Critics point out that these macroeconomic policies pursued across the developing world have led to deflationary pressures and have resulted in the erosion of capacities of national governments to intervene to pursue independent macroeconomic policies. This does not necessarily mean 'end of geography' or 'end of the nation-state' (Scholte 1997).

> Borders still matter for most activities as they did in the past, and the different social and cultural profiles of countries still leave their imprint on the organisational forms of activities as they have always done. What is different today, however, is that the majority of activities, regardless of where or how they continue to be executed, are now constrained in one way or another by the same financial parameters established by the world's capital markets. Although unusual since without historical precedent, the financialisation of the global economy and the homogenisation of priorities is entirely in keeping with capitalism's 500-year history as a commodity system and essentially represents nothing other than a new, defining stage in that history. (Lysandrou 2005: 793)

While at times there has been an over-emphasis on the dichotomy between state and market, careful scrutiny of third world experiences suggests that neo-liberal economic policy has resulted in the reconfiguration of the 'developmental state' as a partisan guardian of corporatist interests. States have intervened very strongly in favour of market forces, often used as a euphemism for big capital even while withdrawing crucial supports to labour and other marginalised sections. The state-versus-market debate has often over-simplified the range of policies that states have pursued to push forward the logic of neo-liberal economic governance.

The changes in production organisation in the age of globally mobile capital have raised newer issues for labour studies (Huws 2006). As against the older Fordist production organisations, the post-Fordist regimes went in for flexible specialisation, with an emphasis on a lean production base that enabled international mobility at an unprecedented scale. Searching for cheapest labour, raw materials, markets, low tax and less stringent environmental regulations, capital could acquire a degree of mobility that was un-precedented. The rise of white collar employment, development of information technology (IT) and IT-enabled services and global integration facilitated this transnational mobility of capital. Although critics have questioned the extent of this mobility and its capacity to transform production conditions across the Third World, nevertheless this capacity to look for cheap labour within and beyond borders has resulted in significant erosion of the bargaining power of labour within many national contexts (Harriss-White 2002). The focus of the book is much narrower. We do not even attempt to study the impact of globalisation on the Indian plantation sector in a systematic manner. Still, the changes that we have studied in the world of work of the plantation labour can be understood only with reference to these larger changes in the global economy, more so because the tea sector in India historically has been more closely integrated with the global economy than many other sectors.

Labour in Globalising India

Pre-reform India is generally described as an over-regulated, inefficient economy. Even in the post-reforms period, labour market reforms have been one of the most controversial issues in economic policy. It has been argued that India's labour market regulations are heavily biased towards the organised sector that employs less than 10 per cent of the total workforce. This rigidity in the labour market, it is argued, has created disincentives for employers and has throttled the expansion of employment in the organised sector. As firms find it difficult to lay off labour they do not employ additional labour during periods of boom as well (Anant 2009). As a response to these labour laws, firms tend to employ more capital-intensive technologies or limit their sizes to avoid the law. Anant et al. (2006), in a survey of the literature on labour market flexibility in India, conclude that 'the case for formal rigidity is weak, and that the growth of value added in the organised sector, along with negligible formal employment, supports the case for a *de facto* flexibility' (Anant 2009: 198). Several studies have documented the role of the unorganised or informal sector in India's recent growth experience. Even while facing challenges from a globally integrated market, India's vast unorganised sector has been playing a significant, though often less-acknowledged, role in India's post-reform growth.[5] Apart from the overwhelming significance of the unorganised sector (Harriss-White and Gooptu 2000), in terms of employment and output that has been so convincingly brought out by the report of the National Commission for Enterprises in the Unorganised Sector (NCEUS) (NCEUS 2007), the growing and changing linkages between the formal and informal sector have significantly been shaped by the emerging challenges faced by the Indian economy in recent decades.

Outlining the significance of the inter-sectoral differences and fragmentation of the labour market in the Indian economy, Mazumdar and Sarkar (2008: 13) have argued that the overall distributional outcomes of the ongoing liberalisation and globalisation would depend on : (*a*) distribution of labour between various segments, and the way it changes in response to the developments in the external sector; (*b*) the distribution and extent of changes in the inter-sectoral wage gaps over time; and (*c*) the changes in the distribution of earnings within each sector. The organised sector itself has gone in for a variety of ways to informalise its operations (Deshpande et al. 2004).[6] There has been a greater degree of outsourcing and subcontracting across a range of industries. This, in fact, has been the case not just in India, but elsewhere too. The informalisation of the

---

[5] Hensman (2001) draws careful distinction between the process of globalisation that increases informalisation and the preponderance of the informal sector in India. Citing Harriss-White and Gooptu's (2000) analysis of the informal economy, she argues that the 'huge preponderance and more rapid growth of the informal sector in comparison to the formal sector is the result neither of globalization nor of lack of flexibility in the latter, but of government policies favouring the former, combined with employer preference for a minimum of regulation, or preferably none at all'.

[6] In their study on labour market flexibility in Indian manufacturing, Deshpande et al. (2004) found that in the post-liberalisation phase, not only did the share of non-permanent workers increase but the share of casual workers in non-permanent workers increased even faster. Significantly, it was the bigger firms that resorted more to the employment of non-permanent workers.

formal economy,[7] as part of the process of economic restructuring in
a globalising India, has several implications for the 'world of work'
(Nayyar 2003). The complex ways of subcontracting, switching over
from permanent to contract labour, and outsourcing across various
regions have created newer forms of work organisation and have
changed the way labour work and live, even in the formal economy.
So far as the agro-food systems are concerned, several important
changes, such as increasing corporate control, vertical integration of
agro-food markets, increase in long distance transfer of food items,
and increasing role of supermarkets in food distribution have been
noticed (von Braun and Diaz-Bonilla 2008). The plantation crops in
general have been at the forefront of many of these changes.

This study, which essentially focuses on the changes in the world
of work for labour employed in one of the oldest 'globally-integrated
sectors' of the Indian economy, provides glimpses to the ways
through which labour is confronting the challenges of globalisation
in India. India's tea sector, however, has its own specificities. On
the one hand, it belongs to the formal/organised sector and labour
legislation aiming specifically at the sector has been at the forefront
of the debates on labour market flexibility in India. On the other
hand, this is a sector that employs a large labour force, has the
highest share of female workers among all the organised industries,
and essentially combines agricultural and industrial processes in
the organisation of its production. It is a sector that has gained from
the growth impetus of the domestic economy, through a rise in
domestic demand and consumption, and has simultaneously faced
the challenges of rising competition because of trade liberalisation, in
global markets and, to a lesser extent, even in the domestic market.
The performance of the tea sector in India is, thus, partly a story of
adjustment within the economy in the face of changing conditions
within and outside the economy. More importantly, the way

---

[7] Hensman (2001: 14) observes: 'Much of the explanation of the slowdown in
growth or even absolute decline of employment in the formal sector lies in the
large-scale transfer of jobs outside this sector. Conversely, much of the growth
in informal sector employment is accounted for by this transfer; therefore *this
is not the creation of new jobs, but simply the downgrading of existing ones*' (emphasis
in original).

such restructuring affects labour brings out the unevenness and inequalities inherent in the processes of 'opening up of opportunities under globalisation'. Increasing casualisation and informalisation has been noticed to be one of the key responses of the Indian industries to the pressures of global competitions. How exactly these challenges are negotiated at the level of household strategies of labour is one of the starting points of this primary research. Here again, the roles of history and geography are more than incidental. History of plantation economies in general, and that of plantations in Assam in particular, have been extensively researched. The singular significance of 'coercion', exercised through a variety of means — ranging from the use of state power through various institutional and non-institutional mechanisms, to use of private means of punishments by planters to 'discipline' labour — was important in the production organisation for a considerable period. The creation of an 'enclave economy' in plantations and that of a 'non-native' labour force through 'indentured labour' have had consequences that went beyond the specific conditions under which the labour force was created.

In spite of the protection of laws and the presence of trade unions in plantations, the tea labour force, largely consisting of the descendants of those who had migrated to colonial Assam as bonded/indentured labour from the tribal belts of Chhota Nagpur region, are among the most vulnerable sections of India's 'organised' labour class. The specific conditions under which they live and work within the plantations that are often isolated from, and poorly connected with, the local economy and society impact upon their work opportunities as well as bargaining power. They are not only marginalised in terms of their position in relation to work and employment, most of them also belong to the Adivasi communities, and yet do not have access to the legal protection provided by the Constitution of India to the Scheduled Tribes elsewhere. The irony of the situation is best captured by the fact that relatives of plantation workers who did not migrate to the plantations are considered Scheduled Tribes and are entitled to the affirmative action policies adopted by the central and state governments, but the descendents of tribal migrants to Assam are not counted as Scheduled Tribes. Although most of the tea labour force in Assam was born and brought up in the gardens, their integration in Assamese society remains weak. Their movement to gain Scheduled Tribe status has failed to draw widespread support in Assam, but has been candidly opposed by many political and ethnic groups. Thus, the tea garden labour in Assam encounter marginality at several levels and these multiple marginalities tend to reinforce each other to create strong impediments in their upward mobility. The study of intergenerational occupational mobility of tea plantation workers in Assam is thus an attempt to unravel some dimensions of this layered experience of multiple exclusions. Since the focus here is on a relatively narrow set of variables, that is, mostly related to the conditions of work and employment, the analysis is far from exhaustive and adequate to understand the 'life-worlds' of the workers, but it provides important insights into the way labour in globalising India struggles to live and work. The 'freedom' of workers to move across occupations and employers is an important dimension of their right as workers. It is important to note here that while mobility of factors of production sans national boundaries is often used as a theoretical argument to highlight the benefits of free trade, labour 'for the most part' is 'completely unfree to move, confronting ever *rising* barriers to economic — let alone forced — migration' Harriss-White (2002: 1; emphasis in original). Needless to add, this in itself is an important aspect of understanding labour under globalisation, or, to be more precise, labour at the margins of globalising India.

The idea of 'labour at the margins' also brings in the question of spatial unevenness in the processes of globalisation. Capitalism, as has been argued eloquently by Lenin, is an uneven process. Colin Leys, for example, remarked in the context of Africa: 'Capitalism does not develop every region similarly. There are poles of growth and margins of decay; areas of high mass consumption and other areas that are mere pools of surplus labour, famine and anarchy' (Leys 1994: 15-16 cited in Harriss-White 2009). Regional convergence has been one of the powerful, motivating and legitimising metaphors of neo-liberal economic policies. Domestic deregulations and trade liberalisation, by unleashing the forces of dynamic growth, have been expected to narrow down inter-regional divergences in economic attainments. Without attempting even to summarise the rich literature on growth divergence and rising inter-regional disparities in India (Ahluwalia 2000, 2002; Singh et al. 2003; Bhattacharya and Sakthivel 2004; Nayyar 2005), it would suffice to note that the fate of regionally concentrated industries that have played a prominent role in the past is also an important window to understand the regionally embedded growth trajectories in India. The relatively poor performance of Assam and the other states of India in terms of economic development is often explained away by referring to the circular logic of backwardness fuelling political instability, which in turn discourages inflow of investment and thus perpetuates backwardness. Turning the gaze away from this highly polarised debate, this study seeks to examine the case of a sector that had brought capital and labour from outside and contributed significantly to the country's export earnings and yet failed to create dynamic growth linkages within the local economy, notwithstanding the recent growth of small-scale tea plantations in Assam. What happens to labour in this important sector in Assam and particularly the reasons behind their occupational mobility, we would like to argue, is an important way to make sense of the development dynamics of the region.

Situating Labour in the Context of the Present Crisis

The tea industry in India is said to be going through a crisis since the early 1990s, primarily because of a fall in tea auction prices. Other manifestations of this crisis include decline in exports, closure and abandonment of tea gardens, increasing labour unrest at times leading to violent protests and confrontations, non-payment and curtailment of wages and other statutory benefits of workers, declining living standards and worsening human security in the tea gardens. There is, however, divergence of opinion regarding the precise nature of the 'crises'. For example, it has been pointed out that the focus of the discussions on the tea sector is on the marketing front, and not so much on production. Tea production in India is not only enough to meet the domestic demand but also to export around 2 lakh tonnes a year. From the producers' point of view the major question is securing a remunerative price for tea, particu-larly after the opening up of imports under the WTO Agreement (*This Hindu*, 2010). Although imports for domestic consumption are still negligible, this cannot be a ground for much complacency as the fairly high prevalent import tariff could be reduced gradually to the global level of

3 to 4 per cent in course of time (Ramadurai 2002). Tea gardens in India experienced closures and abandonment especially during the years 2002-4 in the Dooars region of West Bengal, in Ponmudi, Trivandum District and Peermade, Idukki District of Kerala with a lesser magnitude in Assam and Tamil Nadu. At present, there are reports of continuous closures in some of the tea gardens in West Bengal and Kerala (Talwar et al. 2005).

There have been reports of strikes, unrest and even cases of violent confrontations between workers and management personnel. The crisis in the tea sector, particularly in Assam, has led to labour unrest and apprehensions regarding the future prospects of the sector as well as the workers depending upon the sector (Lahiri 2000; Misra 2003). In Dibrugarh district, a tea garden manager was hacked to death by agitating workers. This event was far from an aberration, rather it was an extreme manifestation of a much deeper crisis affecting the sector (Misra 2003). The relation between tea garden employers and labourers has deteriorated over the past decades, aggravated by the crisis in the tea industry. However, there appears to be a larger dimension to this tension. The exclusion of these communities from the mainstream because of their non-inclusion in the Scheduled Tribes of Assam, despite the fact that they are tribes, is at the bottom of the growing restlessness among the tea garden labour community[8] (Mishra 2007; Gohain 2007). This trend of growing labour unrest is not limited to Assam alone. There have been global concerns about the declining standard of living of

---

[8] On 24 November 2007, when 6,000 Adivasis from the tea gardens of Assam gathered at a protest rally in Guwahati to demand ST status, their rally was attacked by local shopkeepers and youth in retaliation to violent attacks by some of the agitators. A report in *Outlook* magazine describes the incident as follows: 'In a retaliation of a sort never seen on the streets of Guwahati before, the shopkeepers and local residents came down heavily on the adivasis. Within less than an hour of the frenzy, a 46-year-old adivasi protester lay dead, 250 others were wounded. In the melee, an adivasi girl was stripped naked and pursued by miscreants. She was rescued by an elderly man, who gave her his shirt and she reached the safety of a nearby police station. ... Also nursing her wounds at the hospital was Marconi Soren, an 18-year-old adivasi girl from Mornai Tea Estate in the western Kokrajhar district. "Someday, we shall get justice," she said.' See http://www.outlookindia.com/printarticle.aspx?236198

the labourers in the tea gardens (Biswas et al. 2005; Lines 2006). *The Guardian*, for example, in one of its reports claimed that in West Bengal alone, 'at least 700 Indian tea workers have died from diseases linked to malnutrition in the last year after 16 estates were closed' (*The Guardian* 2007). The Centre for Education and Communication (CEC), New Delhi, organised a fact-finding visit by independent experts to investigate the conditions of workers in the tea gardens in 2003. The findings of the team confirmed the non-payment of wages, and in some cases reduction of wages and increase in work load, non-deposit of provident fund and severe curtailment of health and housing facilities (Banerjee et al. 2003).

Tea estates are enclave economies in which the workers are provided with subsidised ration, drinking water, housing, health, and other basic facilities by the management. With the closure and abandoning of the tea estates, the living conditions of the workers have been drastically affected. So far as housing and sanitation are concerned, the conditions were already poor and have simply deteriorated further. A rather alarming report mentions, 'While the children have stopped going to school, the women are migrating to cities in search of illusive jobs. Some others survive by crushing stones or by collecting wild fruits and vegetables. In some cases, impoverished women are forced to sell their bodies as a means of survival'. These problems have compounded with existing social problems — particularly non-compliance to labour legislation that already characterises the industry. There have been reports of malaria epidemics and worsening health conditions and even starvation deaths from the tea gardens (Biswas et al. 2005). Poor food and water has allegedly led to the deaths of over 350 people in four tea estates of Dooars region in six months in the year 2003. Similarly, there are more than 12 cases of starvation deaths and suicides reported from Kerala. Since all the estate hospitals have closed down there is no medical help available to the workers.

Faced with the decline in job opportunities, worsening living conditions and loss of livelihoods, the tea garden workers have started to look for work outside the tea gardens. It is the casual labour force that bears the brunt of downsizing in the gardens. They are the first to face evictions from the tea gardens and for them alternative sources of livelihood are difficult to come by. These casual workers are not covered under the Plantations Labour Act, 1951 (Appendix Ia) or Assam Plantation Rules, 1956, and therefore face severe employment insecurity.[9] Their predicament is all the more severe, as they get work only for five to six months during peak plucking season, after which they are forced to look for employment outside the plantations. As such there has been a gradual process of land alienation from the tea garden labour community, and given their low levels of education, skills and exposure, they find it difficult to get jobs outside the garden.

Tea garden workers and their descendents[10] living outside the gardens face a variety of constraints (Misra 2003, 2007). First, because of the very nature of their settlements the labour community has had noticeably poor access to anti-poverty social security, agricultural extension and other welfare schemes. The selection of beneficiaries is normally biased in favour of the better-off and powerful sections of society, and to the extent that the tea garden labour community is still not 'integrated' into the mainstream local community, and therefore labourers' access to local government remains marginal. Thus, apart from their geographical exclusion, the lower political bargaining strength of the community has resulted in exclusion from grass-roots political bodies such as *gram sabha*s. There have been attempts to mobilise the 'tea tribes' through political and quasi-political organisations, particularly around the demand to be included in the Scheduled Tribes list. Land alienation through the classic 'debt-induced land alienation' route, as well as through occasional use of coercive methods, pushes the Adivasis back to the tea gardens

---

[9] In a recent study based on Darjeeling plantations, Besky (2008) has argued that to the extent that international negotiations and civil society, following the neo-liberal logic, focus primarily on fair trade regulations, undermine the state. These negotiations are too abstract and undermine the local institutional processes that have attempted to protect workers' rights. 'The rapid adoption of fair-trade certification in Darjeeling tea production threatens to erode the power of labor unions, leaving the question of social justice in plantations in the hands of an unaccountable non-state bureaucracy' (ibid.: 7).

[10] The tea garden labourers, including those settled outside the garden are also called Adivasis, tea-tribes or ex-tea garden labour community. These terms are politically loaded statements of self-identification, particularly in Assam. Without taking a stand on this, we would use them interchangeably.

or forces them to migrate to new areas such as Arunachal Pradesh. The tea gardens, on the other hand, are severely downsizing and hence the only option for the tea garden labourers is to join the agricultural labour market at abysmally low wages. Low levels of literacy, high drop-out rates among the children and alcoholism further aggravates their plight. Because of their low fall-back positions and lower access to institutional support from the government and non-government agencies, the tea garden labour community is at times disproportionately affected by natural disasters like floods. Natural disasters have 'amplified the livelihood crisis which these communities face in Assam' (Mishra, 2005).

In the backdrop of the context described above, this study attempts to examine the extent, types and implications of intergenerational occupational mobility among tea garden labourers in Assam. More specifically, the objectives of this research were to investigate:

- The extent of intergenerational occupational mobility and occupational diversification among the tea garden labour communities in Assam;
- the pattern of occupational diversifications among tea garden labour households;
- the constraints and opportunities faced by the tea garden labour households in diversifying their sources of earnings and employments;
- the determinants of occupational mobility and immobility at the household level.

Database and Methodology

The study is based on both secondary and primary data. Data provided in *Tea Statistics* and other published sources was used to analyse the trends in growth of gardens. The data used is (*a*) area under tea and (*b*) production of tea. To understand the trends in productivity and labour use in the tea sector, district level data published by the Tea Board of India was used. To understand the extent and pattern of occupational mobility, household level data, generated by the primary survey, was utilised.

A two-stage sampling technique was followed to select the sample households. At the first stage, depending upon the relative share in total tea production in Assam, three tea producing districts of Assam — Dibrugarh, Sivasagar and Lakhimpur — were selected for the survey. Three tea gardens were randomly selected from amongst the gardens operating in the three districts. The small tea gardens were excluded from the purview of the study. After selection of the gardens, 125 labour households were randomly selected from each of the tea gardens under study. Further, to investigate the determinants of occupational mobility and immobility, 125 ex-tea garden labourer households who have diversified their sources of earnings and employment were also included from the vicinity of the tea gardens. Typically, the tea garden labour households, who have moved out of the tea gardens, live in nearby areas. After selection of these localities, specific ex-tea garden households were selected randomly. Thus, in total 750 households, that is, 375 tea labour households and 375 ex-tea garden labour households were surveyed. Separate structured questionnaires were canvassed in the tea garden labour and ex-tea garden labour households. To supplement this, 150 randomly chosen households (that is, 50 from each of the three gardens under study) were asked about their family tree through semi-structured open-ended interviews in order to map intergenerational mobility. There was a separate questionnaire to gather information on employment and production details of each of the gardens. The quantitative information was supplemented and cross-checked through informal interviews, participant observation and focused group discussions. Separate focused group discussions were held with supervisory staff, permanent and casual workers, women workers, child workers as well as ex-tea garden workers.

The book is organised as follows. The introductory chapter provides a brief discussion on the context, objectives, database and methodology of the study. The historical and contemporary significance of the plantation sector in India is discussed in the second chapter. The growth performance of the tea sector in India is analysed at a disaggregated level in this chapter. The focus of the chapter is on bringing out the temporal changes in the tea gardens of Assam, in terms of growth of area, production and yield of tea at the district level, in a comparative perspective. The changing structure of tea production in the gardens, and its implications for productivity, has also been included in the chapter. The third chapter, primarily relying on secondary data, discusses the changing demand-supply scenario in the labour market in Assam's tea gardens. The changes in employment, labour productivity, patterns of labour use, employment elasticity, as well as the changing composition of the tea labour force, are all examined at a disaggregated level here. All these provide a backdrop to the context in which decisions about occupational choice is made in the tea sector. The fourth chapter deals with the extent, pattern and determinants of intergenerational occupational mobility and diversification of occupations among the tea garden labour households. Relying on the primary data collected from the field survey the subsequent chapter attempts to examine the factors that constrain or facilitate occupational mobility in the specific context of tea garden labour households. Conclusions and policy implications are put forward in the final chapter.

2

The Tea Sector in Assam in a Comparative Perspective

The tea sector has played an important role in the regional economy of Assam and northeast India. Not only does Assam produce around 52 per cent of the tea produced in India, the tea sector is the largest employer in the organised sector in the state. It employs around 590,000 workers directly and provides sustenance to another 660,000 as resident dependents in the tea estates. The performance of the tea sector in Assam needs close examination in a comparative perspective. The historical context of the development of the plantation economy of Assam is discussed briefly here, as one of our key arguments is that much of the contemporary problems in the tea plantations have their roots in the colonial roots of the sector. In this chapter the growth performance and structural features of the tea sector in Assam have been analysed at a disaggregated level and in a comparative perspective. The manifestations of the much-discussed crisis in the tea industry have been investigated to bring out the implications for changing production conditions in the tea gardens of Assam.

The Plantation Sector and World Capitalism

In the global expansion of capitalism, the plantation sector has played a significant role. Plantation signified the integration of land and labour in the colonies to the global production regimes. The historical specificities of the emergence of plantation economies during the colonial period shaped much of its production organisation. It is hardly surprising that plantation economies are generally regarded as 'among the most representative examples of colonial economies' (Dasgupta 1983: 1280). The specific ways that imperial capital and pre-capitalist, subsistence economies of the colonies were brought together to create these largely export-oriented agro-industrial complexes, and the role of the colonial state in the process, have conditioned the way these economies have been managed and developed. 'The history of tea plantations in India reads much like the history of capitalism world over' (Raman 2010: 45). Notwithstanding the colonial legacy, the production conditions in plantations are supposed to have undergone radical changes, particularly in terms of technology, labour management and marketing (Gregor 1965). However, studies also point out that labour relations in plantation economies continue to remain oppressive (Raman 2010) and the inhuman conditions of labour in plantations have been a matter of great concern in recent decades (Khawas 2006; *The Guardian* 2007).

Conceptualising Plantation Economy

The special features of plantation economy have been the focus of much discussion in the literature. Plantation economies typically include characteristics of both agriculture and industry. That has led to a continuing interest in defining the essential features of the plantations as an economic system, as a specific mode of organisation of production.

'A plantation is a capitalist type of agricultural organisation in which a considerable number of unfree labourers were employed under a unified direction and control in the production of a staple crop' (Genovese 1967 cited in Graham and Floering 1984). In a report, The Royal Commission on Labour in India of 1931 defined the plantation system in the following words:

> The plantation system connotes the acquisition of a limited but fairly extensive area for the cultivation of a particular crop, the actual cultivation being done under the direct supervision of a manager, who in some cases may himself be the actual proprietor. A considerable number of persons (the number may run as high as 4000) are employed under his control in the same way as the factory workers are under the control of the factory manager, but there is one important difference in that the work is essentially agricultural and is not concentrated in a large building. (Fay 1936)

The International Labour Committee of Work on Plantations (1950) defined plantations as

...large-scale agricultural units developing certain agricultural resources of tropical countries in accordance with the methods of Western industry...The institution is essentially a large-scale enterprise, depending on large capital investment, a large supply of labour, extensive land areas, well-developed management and specialisation in production for the purpose of export.

Beckford (1972), quoted in Graham and Floering (1984), pointed out the following:

[Plantations] differ from other kinds of farm in the way in which the factors of production, primarily management and labour are combined. The plantation substitutes supervision — supervisory and administrative skills — for skilled adaptive labour, combining the supervision with labour whose principal skill it is to follow orders.

Graham and Floering (1984: 15), emphasising the role of managerial control in the plantations, argued that a 'modern' plantation estate is

...[an] organisation for the large scale production of commodity crops by a uniform system of planting, cultivation and often on-site processing under a central management and with a trained labour force, sometimes living in estate housing in an environment controlled by the same management. Its foundation is the expert direction and training of its workforce by the use of a technology of detailed routine working and supervision.

In such a perspective, managerial control over labour force is viewed as an essential feature of the plantation system. However, it is the specific nature of the crop that has also been identified as the defining feature of the plantation system. Tiffen and Mortimore (1990: 8) have outlined a number of specific characteristics of plantation crops. These are:

(a) they are tropical or sub-tropical crops having an export market (or in more recent times internal markets); (b) most of them need prompt processing; (c) the crop is funnelled through a single or a few intermediate marketing points for the purpose of bulking, processing or standardisation before reaching the consumer and these properties, generally but not necessarily, provides the scope for vertical integration of production and marketing functions and also facilitates the taxation of the crop; (d) they typically require large amount of fixed capital investment; (e) they generate some activity for most of the year so that the large labour force could be engaged throughout the year; (f) mono-cropping is the characteristic, though not universal, since it makes possible standardised management practices and marketing channels; (g) because of the above mentioned factors, rapid change in the product and processes in such plantations becomes difficult and hence such crops are vulnerable to changes in commodity and factor prices.

However, many of the 'plantation crops' are cultivated not only in large estates but also by small farmers. The growth of small tea growers in India in the past few decades is a case in point. In fact, in 1982, the International Labour Organization (ILO) revised the 1958 Plantation Convention to exclude holdings of less than 5 ha and 10 workers. A plantation, according to Tiffen and Mortimore (1990), is not simply a large farm. It generally cultivates one, or less frequently two, of a restricted range of crop; it has a higher capital to land ratio and also a large, mostly permanent, labour force.

Baak (1991: 89–90), writing specifically in the context of Asian plantations, provides a more comprehensive conceptualisation of the plantations as 'a particular organisation of commodity production with specific characteristics in the economic, social, political and cultural fields'. The specific features of plantations include 'production for markets, the vertical integration of labour, the dominance of management in all spheres of life, and a strong relationship with the state'.

History of Tea Plantations in India

The tea industry in India began with the founding of the Assam Company in 1839, although the potential for growing tea was discovered earlier, in 1824, by Major Robert Bruce when he came across indigenous tea bushes in Assam. After the cancellation of the monopoly rights of the British East India Company to trade with China in 1833 by the British parliament, its directors decided to explore the possibilities of tea production in Assam, which they had annexed in 1925 (Bose 1954: 1–2). Gradually, Indian tea gained popularity among the British consumers and by 1859 there was a 'mad rush to clear the hillsides of Assam for new gardens' (1974: 29). Since then and through the colonial period, the industry saw continuous expansion and consolidation.

In many ways the development of the plantation sector in Assam was determined by shifting colonial interests. It has been argued that production organisation in plantation economies is characterised by a pattern where capital and management were brought from the imperialist countries, whereas labour and land were procured from the colonies. A study of the processes through which land and labour were made available for these profit-oriented enterprises brings in the involvement of the colonial state in creating conditions for establishment and consolidation of the plantation economies to the forefront. Essentially, these involve significant disruption of the pre-colonial/pre-capitalist economies. As in other parts of the world, the colonial state was deeply implicated in this process of forcible creation of a labour force and a land pool for the plantation sector in India. The overall regulative framework, as well as the administrative structures of the colonial state, created conditions under which large chunks of land were made available to the plantations and it was possible for the planters to bring in labour from the relatively densely populated and poorer regions to create a labour force to clear the forests and work in the plantations.

This process of shift of land-use from subsistence agriculture, practised by the indigenous people of Assam, to export-oriented commercial agriculture in plantations was sudden and swift and '(w)ithin fifty years of its inception, the rate of expansion of plantation acreage far outstripped that of the area under traditional crops' (Dasgupta 1983: 1289). This rapid transformation in land-use was made possible through a variety of means including preferential treatment to planters, sale or lease of land at very cheap rates, concessions in land revenue, dilution of existing provisions to accommodate the concerns of the planters, and wilful negligence of the interests of the local farmers and tribal communities, etc. (ibid.; Siddique 1990; Guha 2006). Examining the colonial land administration policies in Assam during 1839–1914, Dasgupta (1983) has argued that discriminatory land revenue policies under which concessional land revenue was adopted for the planters while the revenue burden of the farmers' cultivation of traditional crops was increased not only constrained the growth of traditional cultivation, but also gradually led to the encroachment upon cultivated land by planters. Such a policy of negligence resulted in gross underutilisation of land in the plantations. In the Brahmaputra valley, in 1860, only 18.57 per cent of land under plantations was actually under tea cultivation and this increased to 31.33 per cent by 1914 (Dasgupta 1983: 1283). On the other hand, the policy led to displacement of the indigenous population from crucial livelihood resources like forests and grazing lands. The enclosures that resulted in the creation of the enclave economy of the plantations had long-term implications for the growth of the local economy. It also set the basis for a severe disjuncture between the advanced, capitalist, export-oriented agriculture of the plantation sector and the subsistence agriculture of the local economy. Behal (2006: 159) has described the changes brought about by the tea plantations in the following words:

> [F]rom the 1870s onwards, the entire geographical landscape of the Assam Valley was transformed. Tens of thousands of acres of jungle and wasteland were converted into private estates, inhabited by labourers, Indian clerical staff, and European managers and their assistants. Through mergers of small gardens, large units averaging 1,200 acres in size had emerged as the typical plantation by the late nineteenth century. Most gardens became physically isolated both by geographical distance and deliberate exclusion by fencing off from urban settlements as well as the surrounding rural society. These huge private estates, with compulsorily resident indentured labour in the coolie lines, provided the milieu for the exercise of virtually unlimited powers by the planters over their workers.

Growth of Tea Production in India

There has been a phenomenal growth of tea gardens in India, particularly since the 1990s, mainly due to increase in the number of small tea gardens in various states such as Assam, Himachal Pradesh, Tamil Nadu and Kerala. This has also altered the regional distribution and concentration of tea gardens. While the south Indian states accounted for nearly 80 per cent of tea gardens in the early 1980s, in 2004 their share has come down to 47 per cent. After the inclusion of small tea gardens in the data on total tea gardens since 1998,[1] the share of Assam has jumped from less than 7 per cent to 30–35 per cent in recent years (see Table 2.1).

---

[1] The data on number of tea gardens, area and production of tea has been taken from *Tea Statistics* (Tea Board 2004). The inclusion of data on small tea gardens since 1998 does create problems of comparability and has important implications for the analysis that follows.

TABLE 2.1
Number of Tea Estates in India: 1980–2004

Year	Assam Total	West Bengal Total	North India Total	South India Total	Total All India	Assam	West Bengal	North India	South India
	Number of Tea Estates					*Percentage to All India*			
1980	777	305	2,559	10,829	13,388	5.80	2.28	19.11	80.89
1981	793	307	2,578	10,849	13,410	5.91	2.29	19.22	80.90
1982	796	307	2,520	10,863	13,435	5.92	2.29	18.76	80.86
1983	802	311	2,583	10,877	13,460	5.96	2.31	19.19	80.81
1984	808	311	2,589	10,879	13,468	6.00	2.31	19.22	80.78
1985	844	323	2,643	10,887	13,537	6.23	2.39	19.52	80.42
1986	844	322	2,645	10,894	13,821	6.11	2.33	19.14	78.82
1987	845	330	2,654	10,901	13,555	6.23	2.43	19.58	80.42
1988	848	336	2,663	10,905	13,568	6.25	2.48	19.63	80.37
1989	848	347	2,957	10,920	13,877	6.11	2.50	21.31	78.69
1990	848	347	2,931	10,928	13,859	6.12	2.50	21.15	78.85
1991	848	347	2,934	10,936	13,870	6.11	2.50	21.15	78.85
1992	851	347	2,949	10,969	13,918	6.11	2.49	21.19	78.81
1993	850	347	2,948	10,988	13,936	6.10	2.49	21.15	78.85
1994	1,012	348	3,141	31,975	35,116	2.88	0.99	8.94	91.06
1995	1,196	343	5,340	31,979	37,319	3.20	0.92	14.31	85.69
1996	2,472	543	6,746	31,959	38,705	6.39	1.40	17.43	82.57
1997	2,472	453	6,749	31,958	38,707	6.39	1.17	17.44	82.56
1998	25,708	1,115	31,155	56,960	88,115	29.18	1.27	35.36	64.64
1999	30,942	1,451	36,836	62,031	98,867	31.30	1.47	37.26	62.74
2000	39,151	1,540	45,202	66,808	112,010	34.95	1.37	40.36	59.64
2001	40,795	1,554	48,261	68,398	116,659	34.97	1.33	41.37	58.63
2002	40,795	1,554	48,261	68,398	116,659	34.97	1.33	41.37	58.63
2003	43,293	8,709	60,629	68,398	129,027	33.55	6.75	46.99	53.01
2004	43,293	8,709	60,629	68,398	129,027	33.55	6.75	46.99	53.01

Source: Tea Statistics (various years). See Tea Board (2004)

Tea production in India, which started during the colonial period, expanded rapidly since 1870. The British East India Company, and later the colonial government, supported the expansion of the tea plantations in Assam and elsewhere through a series of legislative and institutional support (Behal and Mohapatra 1992; Das Gupta 1992a; Siddique 1995; Evans 1995). During this period, there was steady growth in the area under tea plantation; there were many improvements in tea production technology; significant gains in productivity were also achieved (see Table 2.2), although there were periods of crises because of war and political instability (Karmakar and Banerjee 2005).

TABLE 2.2
Growth of Tea Industry during Colonial Period

Year	Area under Tea (in '000 ha)	Production (in million kg)	Average Yield (in kg/ha)
1850	0.75	0.1	130
1890	152	57	373
1918	275	173	692
1939	337	205	610
1947	309	255	822

Source: Tea Statistics (various years). See Tea Board (2004).

During the post-Independence decades, area, production and yield of tea have registered a steady increase. Taking a three-yearly moving average, it is found that production and yield have suffered a serious setback since 1998, although the area under tea has expanded due to increase in small tea gardens (see Figure 2.1). The growth performance of tea in India comes out in clearer terms, when the variations in growth performance among decades are taken into account. While for the period 1950–2004 the growth rate of production has been close to 2.4 per cent, the area under tea has expanded at a rate of less than 1 per cent per annum. Thus, the major contribution to growth in production has come from rise in productivity. It is important to note that the highest increase in production as well as yield was recorded in the 1970s. While the highest growth rate in area under tea has been registered during the 1990s, this is the decade when production growth was at its lowest and yield also shows negative growth (see Table 2.4). Although the trends in the growth of area, production and yield could have been influenced by the inclusion of small gardens since 1998, even for the period 1998–2004, all these findings — viz., of slow growth of production, negative growth of yield and comparatively high growth rate of area — also hold good.

Tea in India: A Global Perspective

Since India has been one of the major global players in tea production, consumption and exports, the changing trends in global tea production have important implications for the Indian tea sector (see, for example, Mitra 1991; Rao and Hone 1974). An attempt has been made here to selectively review the changing position of India

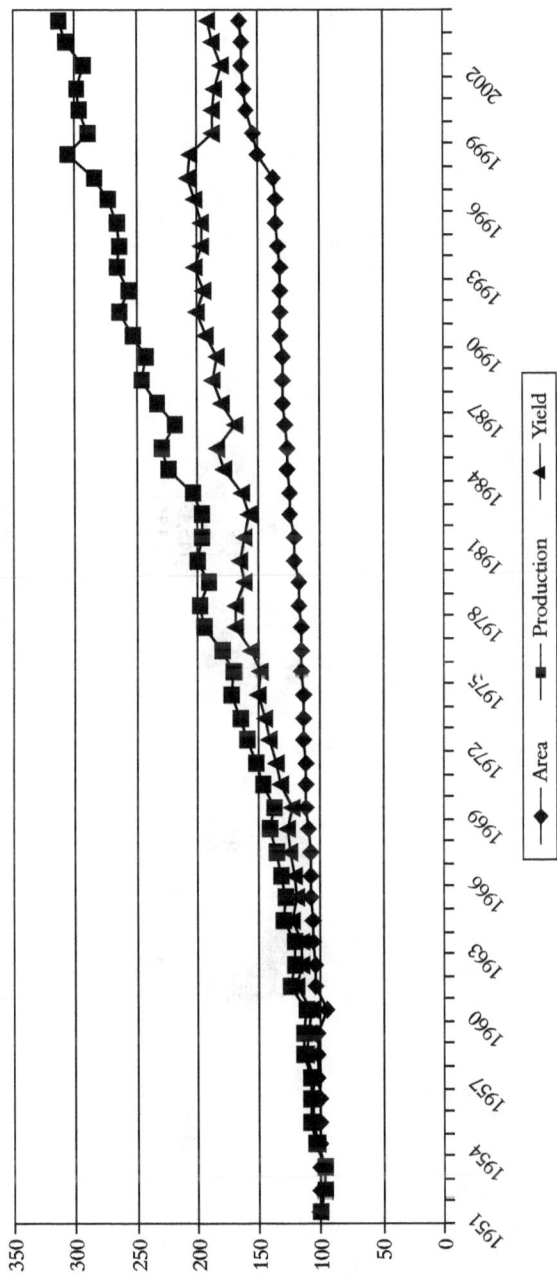

FIGURE 2.1
Index Number of Area, Production and Yield of Tea in India: 1950–2004

Note: Based on data provided in Table 2.3.

TABLE 2.3
Area, Production and Yield of Tea in India: 1950–2004

Year	Area	Production	Yield	Three Years moving Average			Index (Base=1951)		
				Area	Production	Yield	Area	Production	Yield
1950	315,656	278,212	881						
1951	316,870	285,399	901	105,623	95,133	300	100	100	100
1952	317,916	278,671	877	105,972	92,890	292	100	98	97
1953	318,642	278,777	875	106,214	92,926	292	101	98	97
1954	319,478	295,519	925	106,493	98,506	308	101	104	103
1955	320,238	307,704	961	106,746	102,568	320	101	108	107
1956	320,588	308,719	963	106,863	102,906	321	101	108	107
1957	323,285	310,802	961	107,762	103,601	320	102	109	107
1958	325,357	325,225	1000	108,452	108,408	333	103	114	111
1959	326,924	325,955	997	108,975	108,652	332	103	114	111
1960	300,738	321,077	971	100,246	107,026	324	95	113	108
1961	331,229	354,397	1,070	110,410	118,132	357	105	124	119
1962	332,524	345,735	1,040	110,841	115,245	347	105	121	115
1963	334,036	346,413	1,037	111,345	115,471	346	105	121	115
1964	337,874	372,485	1,102	112,625	124,162	367	107	131	122
1965	341,762	366,374	1,072	113,921	122,125	357	108	128	119
1966	345,256	375,983	1,089	115,085	125,328	363	109	132	121
1967	342,653	384,759	1,123	114,218	128,253	374	108	135	125
1968	351,065	402,489	1,146	117,022	134,163	382	111	141	127
1969	353,359	393,588	1,114	117,786	131,196	371	112	138	124
1970	354,133	418,517	1,182	118,044	139,506	394	112	147	131
1971	356,516	435,468	1,221	118,839	145,156	407	113	153	136
1972	358,675	455,996	1,271	119,558	151,999	424	113	160	141
1973	360,108	471,952	1,311	120,036	157,317	437	114	165	146
1974	361,663	489,475	1,353	120,554	163,158	451	114	172	150
1975	363,303	487,137	1,341	121,101	162,379	447	115	171	149

Year								
1976	364,275	511,817	1,405	121,425	170,606	468	115	179	156
1977	366,276	556,267	1,519	122,092	185,422	506	116	195	169
1978	369,184	563,846	1,527	123,061	187,949	509	117	198	169
1979	372,980	543,776	1,458	124,327	181,259	486	118	191	162
1980	381,086	569,172	1,494	127,029	189,724	498	120	199	166
1981	383,629	560,427	1,461	127,876	186,809	487	121	196	162
1982	394,170	560,562	1,422	131,390	186,854	474	124	196	158
1983	396,066	581,484	1,468	132,022	193,828	489	125	204	163
1984	398,453	639,864	1,606	132,818	213,288	535	126	224	178
1985	398,966	656,162	1,645	132,989	218,721	548	126	230	183
1986	407,647	620,803	1,523	135,882	206,934	508	129	218	169
1987	411,335	665,251	1,617	137,112	221,750	539	130	233	179
1988	414,347	700,014	1,689	138,116	233,338	563	131	245	187
1989	414,953	688,105	1,658	138,318	229,368	553	131	241	184
1900	416,269	720,338	1,730	138,756	240,113	577	131	252	192
1991	420,470	754,192	1,794	140,157	251,397	598	133	264	199
1992	420,289	732,322	1,742	140,096	244,107	581	133	257	193
1993	418,363	760,826	1,819	139,454	253,609	606	132	267	202
1994	425,966	752,895	1,768	141,989	250,965	589	134	264	196
1995	427,065	756,016	1,770	142,355	252,005	590	135	265	196
1996	431,204	780,140	1,809	143,735	260,047	603	136	273	201
1997	434,294	810,031	1,865	144,765	270,010	622	137	284	207
1998	474,027	874,108	1,844	158,009	291,369	615	150	306	205
1999	490,200	825,935	1,685	163,400	275,312	562	155	289	187
2000	504,366	846,922	1,679	168,122	282,307	560	159	297	186
2001	509,806	853,923	1,675	169,935	284,641	558	161	299	186
2002	515,832	838,474	1,625	171,944	279,491	542	163	294	180
2003	519,598	878,129	1,690	173,199	292,710	563	164	308	188
2004	521,403	892,965	1,713	173,801	297,655	571	165	313	190

Source: Calculated from data provided in *Tea Statistics* (Tea Board 2004).

TABLE 2.4
Growth Rates of Area, Production and Yield of Tea: All India

Period	Growth Rate		
	Area	*Production*	*Yield*
1951–60	0.44	1.89	1.44
1961–70	0.81	2.01	1.89
1971–80	0.63	3.06	2.41
1981–90	0.88	2.88	1.98
1991–2000	2.09	1.73	−0.35
1991–2003	2.20	1.49	−0.70
1950–2003	0.87	2.39	1.51
1998–2003	1.53	0.63	−0.88

Note: Growth rates are compound growth rates.
Source: Calculated from data provided in *Tea Statistics* (Tea Board 2004).

among the major tea producing and exporting countries of the world. Among tropical and subtropical crops, international trade in tea is among the largest in value terms. The historical trends suggest that world tea exports have been worth between one-half and one-third of world coffee exports in value terms and more than that of cocoa, although in recent years changes in international prices have increased the value of cocoa exports to a level higher than that of tea (Lines 2006). However, the structure of tea production and consumption has some distinct characteristics. The major producers of tea are also among the largest consumers. Very few countries produce tea, and export is dominated by fewer countries. Lines (2006) argues that the relative negligence of tea in development circles could be because of the following: (*a*) little tea is produced in the poorest countries and (*b*) the narrowness of the tea market on both production and consumption sides.

The shares of major tea producing countries in world tea production, presented in Table 2.5, shows China with the highest share in world production of tea, around one-third, in 2008. It is a remarkable increase from around 10 per cent in 1961. India was the largest producer of tea with a 36 per cent share during 1961 and maintained this status until 2004; since then India's share in world total tea production has declined continuously. It came down to 21 per cent in 2008. Similarly, Sri Lanka's share in total world production has shrunk from 21 per cent in 1961 to 8 per cent in 2008. Other countries whose share has fallen are Bangladesh, Japan and

TABLE 2.5
Annual Production of Tea in Selected Countries: 1961–2008

(in metric tons)

Country	1961	1971	1981	1991	2001	2004	2008
1	2	3	4	5	6	7	8
China	97,064	179,984	368,223	562,961	721,536	855,422	1,275,384
	(9.87)	(13.76)	(19.52)	(21.98)	(23.49)	(24.96)	(32.75)
India	354,397	435,468	559,583	720,300	847,000	878,000	805,180
	(36.02)	(33.28)	(29.67)	(28.13)	(27.58)	(25.62)	(20.68)
Kenya	12,641	36,290	90,941	203,588	294,620	308,090	345,800
	(1.28)	(2.77)	(4.82)	(7.95)	(9.59)	(8.99)	(8.88)
Sri Lanka	206,488	217,773	210,148	240,747	295,090	324,600	318,700
	(20.99)	(16.64)	(11.14)	(9.40)	(9.61)	(9.47)	(8.18)
Turkey	5,450	33,585	42,606	136,887	142,900	171,200	198,046
	(0.55)	(2.57)	(2.26)	(5.34)	(4.65)	(4.99)	(5.09)
Vietnam	7,500	15,500	21,178	33,100	75,700	100,700	174,900
	(0.76)	(1.18)	(1.12)	(1.29)	(2.46)	(2.94)	(4.49)
Indonesia	77,100	60,922	109,135	139,520	163,068	201,663	150,851
	(7.84)	(4.66)	(5.79)	(5.45)	(5.31)	(5.88)	(3.87)
Japan	81,527	93,111	102,300	87,800	85,000	119,500	96,500
	(8.29)	(7.12)	(5.42)	(3.43)	(2.77)	(3.49)	(2.48)
Argentina	6,486	29,900	22,785	46,075	71,117	70,389	76,000
	(0.66)	(2.29)	(1.21)	(1.80)	(2.32)	(2.05)	(1.95)
Bangladesh	26,542	12,007	38,772	45,012	52,000	57,580	59,000
	(2.70)	(0.92)	(2.06)	(1.76)	(1.69)	(1.68)	(1.52)
Malawi	14,288	18,597	31,965	40,501	36,800	40,250	48,140
	(1.45)	(1.42)	(1.69)	(1.58)	(1.20)	(1.17)	(1.24)
Uganda	5,100	18,000	1,700	8,877	32,857	35,706	42,808
	(0.52)	(1.38)	(0.09)	(0.35)	(1.07)	(1.04)	(1.10)
Iran	10,922	16,000	33,100	42,091	51,160	50,090	42,348
	(1.11)	(1.22)	(1.76)	(1.64)	(1.67)	(1.46)	(1.09)
World	983,785	1,308,424	1,885,907	2,561,050	3,071,611	3,427,455	3,894,029
	(100)	(100)	(100)	(100)	(100)	(100)	(100)

Note: Figures in parentheses show percentage share in total.
Source: FAOSTAT

Indonesia. On the other hand, besides China, countries whose share in total world production has increased are Kenya, Turkey, Vietnam and Uganda.

Similar changes are also noticeable in the share of different tea producing countries in total area under tea. By 2004, China's share in tea went up to nearly 46 per cent, while that of India had come down to 19 per cent. India, which had 34 per cent of total area under tea in the world, has steadily lost its position and by the early 1990s its share had already come down to 19 per cent. While Sri Lanka's share has gone down more sharply, that of Kenya has increased during the period 1961–2008 (see Table 2.6).

TABLE 2.6
Area under Tea in different Producing Countries: 1961–2004

Name of the Countries	Year					
	1961	*1971*	*1981*	*1991*	*2001*	*2004*
China		303,000 (27.74)		1,060,530 (48.90)	1,140,700 (43.24)	1,262,310 (45.49)
India	332,524 (33.69)	356,516 (25.19)	384,242 (34.44)	420,500 (19.39)	509,806 (19.32)	521,403 (18.79)
Sri Lanka	237,713 (24.09)	24,1667 (17.08)	244,919 (21.95)	221,691 (10.22)	188,971 (7.16)	188,720 (6.80)
Indonesia	72,441 (7.34)	98,027 (6.93)	70,274 (6.30)	137,055 (6.32)	150,938 (5.72)	142,782 (5.15)
Kenya	17,756 (1.80)	43,836 (3.10)	78,920 (7.07)	100,629 (4.64)	131,581 (4.99)	131,418 (4.74)
Vietnam	9,140 (0.93)	84,50 (0.60)	N.A.	N.A.	82,000 (3.11)	90,000 (3.24)
Turkey	1,4967 (1.52)	28,351 (2.00)	52,167 (4.68)		76,653 (2.91)	77,000 (2.77)
Bangladesh	31,688 (3.21)	42,637 (3.01)	44,472 (3.99)	47,677 (2.20)	49,313 (1.87)	50,400 (1.82)
Japan	48,790 (4.94)	53,900 (3.81)	61,000 (5.47)	57,600 (2.66)	50,100 (1.90)	50,000 (1.80)
USSR/CIS	63,400 (6.42)	75,500 (5.34)	78,200 (7.01)	N.A.	40,600 (1.54)	41,800 (1.51)
Argentina	29,700 (3.01)	34,300 (2.42)	N.A.	N.A.	36,000 (1.36)	36,300 (1.31)
Iran	20,400 (2.07)	N.A.	N.A.	N.A.	34,664 (1.31)	34,000 (1.23)
Tanzania	7,797 (0.79)	12,403 (0.88)	N.A.	19,415 (0.90)	21,371 (0.81)	22,287 (0.80)
Uganda	7,659 (0.78)	18,508 (1.31)	20,950 (1.88)	20,905 (0.96)	20,870 (0.79)	21,720 (0.78)
Malawi	12,037 (1.22)	15,455 (1.09)	18,423 (1.65)	18,300 (0.84)	18,761 (0.71)	18,663 (0.67)
Taiwan	47,632 (4.83)	34,200 (2.42)	29,000 (2.60)	23,863 (1.10)	18,938 (0.72)	18,600 (0.67)
Grand Total	986,914 (100)	1,112,155 (100)	1,115,811 (100)	1,108,380 (100)	1,497,469 (100)	1,512,487 (100)

Note: Figures in parentheses show percentage share in total.
Source: Tea Statistics of India (Tea Board, various issues).

In terms of exports Kenya has seen a remarkable increase in its share in global tea exports — from 2.02 per cent in 1961 to 21 per cent in 2008. India's share has declined from nearly 35 per cent to 11 per cent during the same period. India held the largest share in global tea exports in 1961 but came down to fourth position in 2008, after Kenya, Sri Lanka and China. In fact, India's tea export has declined in absolute terms as well. The shares of China, Vietnam and Indonesia have increased but that of Sri Lanka and Bangladesh have declined during the period 1961–2008 (Table 2.7). Broadly, a similar trend is noticed in exports in value terms (Table 2.8). India's share in global tea exports has come down from 38 per cent to 11 per cent. Although Sri Lanka's share has also suffered a decline — 34 per cent to 23 per cent during that period — it has retained the top position in 2008 among the tea exporting countries. Kenya and China have occupied the second and third positions in export share in value terms. Both these countries have improved their share remarkably since 1961.

The share of tea exports in total agricultural exports of major tea exporters has been presented in Table 2.9. Sri Lanka, which already enjoyed a high share of its agricultural exports from tea in 1961, continues to have 60 per cent of its agricultural trade from tea. The share of tea in total annual exports from India was 47 per cent in 1961 and by 1991 it had come down to nearly 17 per cent. However, since the 1990s, broadly coinciding with the post-reforms period in India, the share of tea in India's agricultural exports has come down to 2–3 per cent. Partly, this could be because of diversification of India's agro-export basket, but part of the reason could also be the dismal performance of India's tea sector in recent decades. Kenya comes out as an exception where the share of tea in total agricultural exports has increased substantially. But what is important to note is that, as a share of global agricultural exports, the share of tea has come down from nearly 2 per cent to 0.52 per cent during 1961–2008.

Among the major importers of tea are USSR/Commonwealth of Independent States (CIS), United Kingdom (UK) and United States (see Table 2.10). In 2008, Russia accounted for 11 per cent of global tea imports, followed by UK (9.41 per cent), United States (6.97 per cent), United Arab Emirates (UAE) (6.54 per cent) and Egypt (6.42 per cent). It is important to note that the share of UK, the largest importer of tea in 1961 (42.53 per cent) has gone down substantially to 11–10 per cent since the 1990s.

TABLE 2.7

Annual Exports of Tea from Selected Countries (by volume, tons): 1961–2008

Country					Year			
	1961	1971	1981	1991	2001	2003	2004	2008
1	2	3	4	5	6	7	8	9
Kenya	11,957	41,688	84,095	175,625	207,244	293,751	284,309	396,641
	(2.02)	(5.43)	(8.83)	(14.68)	(14.28)	(19.34)	(17.62)	(20.92)
Sri Lanka	193,103	200,800	183,362	212,017	293,524	297,003	298,909	318,329
	(32.61)	(26.14)	(19.26)	(17.73)	(20.23)	(19.56)	(18.52)	(16.79)
China	47,403	71,355	99,267	190,188	252,204	262,663	282,643	299,789
	(8.01)	(9.29)	(10.43)	(15.90)	(17.38)	(17.30)	(17.52)	(15.81)
India	204,959	199,561	238,750	215,144	177,603	174,246	174,728	203,207
	(34.61)	(25.98)	(25.08)	(17.99)	(17.24)	(11.47)	(10.83)	(10.72)
Indonesia	32,242	44,803	71,259	110,207	99,797	88,176	98,572	158,959
	(5.44)	(5.83)	(7.49)	(9.21)	(6.88)	(5.81)	(6.11)	(8.38)
Vietnam	2,266	1,784	8,100	7,953	67,900	58,600	99,400	147,300
	(0.38)	(0.23)	(0.85)	(0.66)	(4.68)	(3.86)	(6.16)	(7.77)
Argentina	5,352	22,454	25,872	36,029	58,110	59,062	67,819	63,512
	(0.90)	(2.92)	(2.72)	(3.01)	(4.00)	(3.89)	(4.20)	(3.35)

Uganda	4,215	15,268	501	7,018	18,220	8,071	36,856	47,222
	(0.71)	(1.99)	(0.05)	(0.59)	(1.26)	(0.53)	(2.28)	(2.49)
Malawi	13,000	18,157	31,017	37,047	36,587	36,924	32,672	36,861
	(2.20)	(2.36)	(3.26)	(3.10)	(2.52)	(2.43)	(2.02)	(1.94)
Bangladesh	16,310	9,000	31,786	26,862	6,400	7,348	10,635	8,259
	(2.75)	(1.17)	(3.34)	(2.25)	(0.44)	(0.48)	(0.66)	(0.44)
Iran	367	644	2,889	0	10,004	7,402	9,491	9,154
	(0.06)	(0.08)	(0.30)	(0.00)	(0.69)	(0.49)	(0.59)	(0.48)
Turkey	0	17,497	3,315	2,314	4,817	6,711	5,929	3,191
	(0.00)	(2.28)	(0.35)	(0.19)	(0.33)	(0.44)	(0.37)	(0.17)
Japan	7,999	1,420	2,708	289	625	845	923	1,767
	(1.35)	(0.18)	(0.28)	(0.02)	(0.04)	(0.06)	(0.06)	(0.09)
World	592,167	768,216	951,994	1,196,095	1,451,098	1,518,571	1,613,633	1,895,807
	(100)	(100)	(100)	(100)	(100)	(100)	(100)	(100)

Note: Figures in parentheses show percentage share in total.
Source: FAOSTAT

TABLE 2.8
Annual Exports of Tea from Selected Countries (by value, US$'000): 1961–2008

Country	Year								
	1961	1971	1981	1991	2001	2003	2004	2008	
1	2	3	4	5	6	7	8	9	
Sri Lanka	233,933	192,142	335,089	426,085	679,951	672,517	732,521	1,258,700	
	(34.26)	(27.64)	(18.48)	(17.29)	(24.10)	(23.01)	(22.39)	(22.80)	
Kenya	12,000	39,001	149,395	274,863	448,677	481,485	463,726	934,921	
	(1.76)	(5.61)	(8.24)	(11.15)	(15.90)	(16.47)	(14.17)	(16.94)	
China	39,128	53,574	210,642	395,441	358,348	385,158	453,672	700,623	
	(5.73)	(7.71)	(11.61)	(16.05)	(12.70)	(13.18)	(13.87)	(12.69)	
India	259,205	201,794	506,832	490,292	367,207	333,408	377,742	590,226	
	(37.96)	(29.03)	(27.95)	(19.90)	(13.01)	(11.41)	(11.55)	(10.69)	
Indonesia	25,700	28,881	100,837	143,122	99,967	95,816	116,018	158,959	
	(3.76)	(4.16)	(5.56)	(5.81)	(3.54)	(3.28)	(3.55)	(2.88)	
Vietnam	2,178	1,790	7,400	9,221	78,406	58,392	95,550	147,300	
	(0.32)	(0.26)	(0.41)	(0.37)	(2.78)	(2.00)	(2.92)	(2.67)	
Argentina	3,824	11,736	25,660	27,661	41,981	34,119	40,512	63,512	
	(0.56)	(1.69)	(1.41)	(1.12)	(1.49)	(1.17)	(1.24)	(1.15)	

Uganda	4,507 (0.66)	13,400 (1.93)	300 (0.02)	6,780 (0.28)	16,230 (0.58)	8,259 (0.28)	37,256 (1.14)	47,222 (0.86)
Malawi	11,500 (1.68)	14,381 (2.07)	34,188 (1.89)	35,895 (1.46)	34,088 (1.21)	42,069 (1.44)	39,360 (1.20)	36,861 (0.67)
Japan	4,708 (0.69)	1,110 (0.16)	3,091 (0.17)	3,660 (0.15)	9,972 (0.35)	13,763 (0.47)	16,732 (0.51)	33,474 (0.61)
Bangladesh	23,239 (3.40)	4,300 (0.62)	40,969 (2.26)	43,088 (1.75)	6,500 (0.23)	8,223 (0.28)	12,168 (0.37)	10,771 (0.20)
Turkey	0	3,873 (0.56)	7,161 (0.39)	3,090 (0.13)	4,073 (0.14)	6,959 (0.24)	6,857 (0.21)	11,232 (0.20)
Iran	69 (0.01)	289 (0.04)	1,693 (0.09)	0	6,254 (0.22)	6,550 (0.22)	8,257 (0.25)	9,546 (0.17)
World	682,754	695,083	1,813,617	2,464,398	2,821,874	2,922,601	3,271,661	5,520,560

Note: Figures in parentheses show percentage share in total

Source: FAOSTAT

TABLE 2.9
Percentage Share in Total Agricultural Trade of Annual Exports of
Tea of Selected Countries (by value, US$'000): 1961–2008

Countries	Year						
	1961	*1971*	*1981*	*1991*	*2001*	*2004*	*2008*
1	*2*	*3*	*4*	*5*	*6*	*7*	*8*
Sri Lanka	67.50	62.01	52.08	64.40	71.39	34.46	59.21
Kenya	12.54	23.16	24.29	42.91	42.85	17.38	35.03
Uganda	4.08	6.06	0.12	4.07	11.62	4.24	5.38
Malawi	40.66	22.40	14.32	7.78	8.93	5.12	4.80
Bangladesh	11.43	3.72	24.46	29.54	6.92	5.15	4.56
India	47.04	28.69	18.79	17.53	7.02	2.18	3.41
China	10.27	3.98	4.70	3.40	2.76	1.50	2.32
Vietnam	3.00	6.94	7.99	1.49	3.87	1.24	1.92
Japan	2.85	0.32	0.29	0.28	0.40	0.61	1.22
Iran	0.11	0.18	1.43	0.00	0.58	0.58	0.67
Indonesia	5.52	5.86	5.39	4.58	2.29	0.42	0.57
Argentina	0.42	0.81	0.40	0.39	0.38	0.11	0.18
Turkey	0.00	0.73	0.28	0.08	0.10	0.06	0.11
World	2.13	1.24	0.78	0.75	0.68	0.31	0.52

Note: Figures in parentheses show percentage share in total.
Source: FAOSTAT

Over the past decades the direction of tea imports from India has also changed. The quantum and its percentage share in total tea exports from India to major countries are shown in Table 2.11 and Figure 2.2 respectively. A major part of total tea exports from India has been to USSR/CIS. It is clear that although India's tea export market is said to have suffered substantially as a result of the disintegration of the former USSR, even in 2008 CIS countries accounted for nearly 45 per cent of India's tea exports. Iraq and UAE are the other countries to whom India has increased its exports. India has also increased its exports to countries in the Organisation for Economic Cooperation and Development (OECD) marginally. Tea exports to UK and members of the Economic Cooperation Organization (ECO) have declined substantially. Share of exports to other countries, comprising countries of Africa, East Europe and Gulf, has declined.

Figure 2.3 depicts the production, internal consumption and exports of tea in India, over the period 1971 to 2002. Both production and internal consumption of tea has increased over the period. But exports have declined marginally from the level where production was evenly distributed between consumption and exports in 1971.

TABLE 2.10

Annual Net Imports of Tea to Selected Countries (volume, in tons): 1961–2008

Country	1961	1971	1981	1991	2001	2003	2004	2008
1	2	3	4	5	6	7	8	9
Russia	–	–	–	–	154,448 (11.15)	168,894 (12.43)	172,145 (12.26)	181,859 (10.85)
USSR	14,900 (2.51)	42,600 (5.72)	84,521 (9.57)	170,000 (14.95)	–	–	–	–
United Kingdom	251,999 (42.53)	226,289 (30.37)	160,408 (18.16)	178,148 (15.66)	164,016 (11.84)	156,636 (11.52)	156,311 (11.13)	157,593 (9.41)
United States	49,594 (8.37)	79,584 (10.68)	86,297 (9.77)	84,332 (7.41)	96,668 (6.98)	94,174 (6.93)	99,484 (7.08)	116,746 (6.97)
UAE	0	2,300 (0.31)	15,079 (1.71)	22,401 (1.97)	40,000 (2.89)	35,923 (2.64)	65,826 (4.69)	109,575 (6.54)
Egypt	22,764 (3.84)	11,002 (1.48)	30,131 (3.41)	88,528 (7.78)	56,403 (4.07)	38,318 (2.82)	2,627 (0.19)	107,586 (6.42)
Pakistan	16,056 (2.71)	31,800 (4.27)	72,531 (8.21)	104,056 (9.15)	106,822 (7.71)	108,147 (7.96)	115,967 (8.26)	100,391 (5.99)
Morocco	13,856 (2.34)	12,855 (1.73)	22,622 (2.56)	24,289 (2.14)	37,701 (2.72)	44,925 (3.31)	45,670 (3.25)	51,872

Year

(Continued...)

(Table 2.10 Continued...)

Country				Year				
	1961	1971	1981	1991	2001	2003	2004	2008
1	2	3	4	5	6	7	8	9
Germany	8,270	11,597	20,052	25,087	37,758	45,787	43,409	50,771
	(1.40)	(1.56)	(2.27)	(2.21)	(2.73)	(3.37)	(3.09)	(3.03)
Japan	1,970	14,072	13,910	36,674	60,396	47,354	56,234	43,116
	(0.33)	(1.89)	(1.57)	(3.22)	(4.36)	(3.48)	(4.00)	(2.57)
Poland	2,285	9,712	24,547	9,523	33,102	30,594	32,119	30,942
	(0.39)	(1.30)	(2.78)	(0.84)	(2.39)	(2.25)	(2.29)	(1.85)
Syria	1,694	2,435	7,812	21,409	22,336	29,036	30,330	26,987
	(0.29)	(0.33)	(0.88)	(1.88)	(1.61)	(2.14)	(2.16)	(1.61)
World	592,466	745,108	883,209	1,137,346	1,385,202	1,359,108	1,404,308	1,675,386

Note: Figures in parentheses show percentage share in total.
Source: FAOSTAT

TABLE 2.11
Major Country-wise Quantum of Total Tea Exports from India
('000 kg): 1951–2004

Country	Year						
	1951	*1961*	*1971*	*1981*	*1991*	*2001*	*2004*
1	2	3	4	5	6	7	8
USSR/CIS	1,028	11,874	40,935	77,807	106,499	82,210	53,392
	(0.50)	(5.76)	(20.26)	(32.25)	(52.80)	(45.02)	(27.01)
OECD*	46,102	29,227	31,602	34,772	25,749	28,309	33,959
	(22.38)	(14.17)	(15.64)	(14.41)	(12.76)	(15.50)	(17.18)
Iraq	673	2,116	5,911	9,017	–	16,864	25,789
	(0.33)	(1.03)	(2.93)	(3.74)		(9.24)	(13.05)
U.A. E.	(a)	(a)	(a)	12,595	7,449	23,349	25,598
				(5.22)	(3.69)	(12.79)	(12.95)
United Kingdom	136,270	123,215	68,852	40,971	23,751	16,102	19,787
	(66.16)	(59.73)	(34.08)	(16.98)	(11.77)	(8.82)	(10.01)
ECO	6,460	7,166	19,105	19,131	14,805	6,382	10,829
	(3.14)	(3.47)	(9.46)	(7.93)	(7.34)	(3.50)	(5.48)
Other Countries	15,450	32,269	34,852	45,234	23,262	9,371	28,314
	(7.50)	(15.64)	(17.25)	(18.75)	(11.53)	(5.13)	(14.32)
Grand Total	205,983	206,292	202,052	241,246	201,720	182,588	197,668
	(100)	(100)	(100)	(100)	(100)	(100)	(100)

Note: i) * indicates other than UK.
ii) ECO: Economic Cooperation Organisation or ECO (Iran, Turkey, Pakistan and Afghanistan are its founding members).
Figures in parentheses show percentage share in total.
Source: Tea Statistics of India (Tea Board, various issues).

Growth of Tea Sector in States of India

The major tea producing states of India are Assam, West Bengal, Tamil Nadu and Kerala. In addition to these, tea is grown in a limited way in Himachal Pradesh, Uttarakhand, Karnataka and Tripura. In recent years, tea has been introduced in some non-traditional areas like Arunachal Pradesh, Manipur, Nagaland, Meghalaya, Mizoram, Sikkim, Odisha and Bihar. The growth performance of tea in these states has been highly uneven. Among all the major tea producing states, the growth rate of area under tea production has been highest in the 1990s.[2] As pointed out earlier, this high growth in area has

---

[2] Mitra (1991: M154) notes that 'for explaining intertemporal movement of aggregate acreage under tea over time, the long-run profitability (indicated as gross profit as percentage of net assets of the tea industry) is found to be the main explanatory variable. The elasticity of acreage with respect to long run profitability is found to be less than unity'.

FIGURE 2.2
Share of Total Tea Exports of India to Major Countries: 1981 and 2001

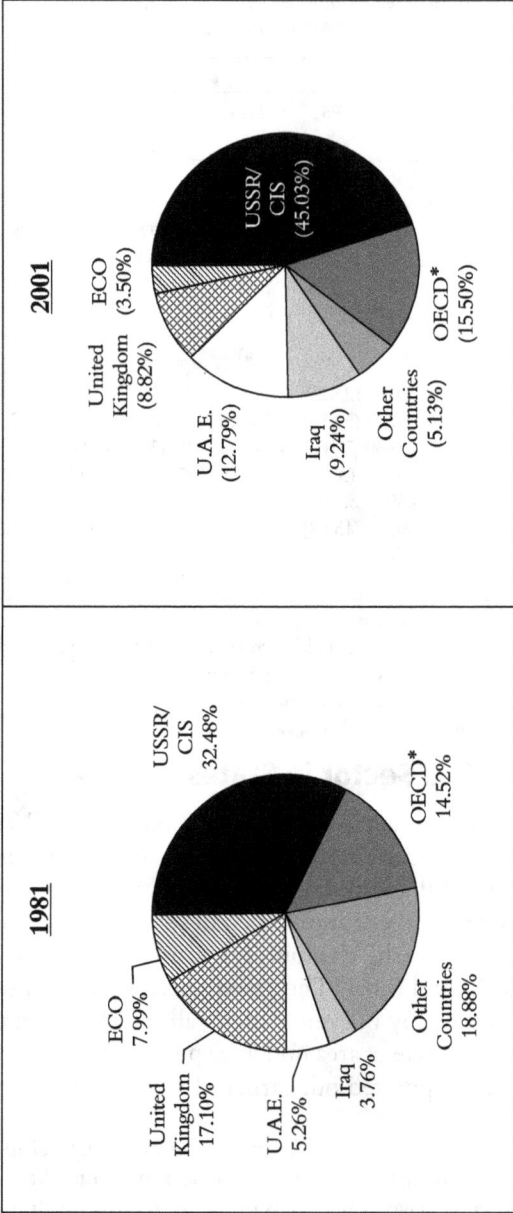

Note: * indicates other than UK.
Source: Tea Statistics of India (Tea Board, various issues).

FIGURE 2.3

Quantum of Tea Produced, Consumed and Exported (in '000 kg): 1971–2001

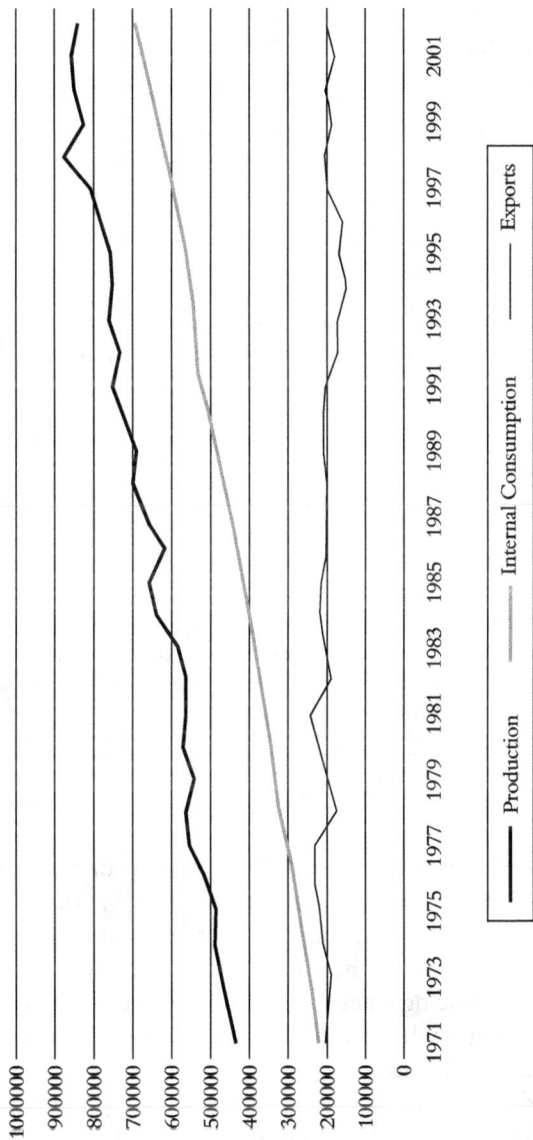

Legend:
——— Production ——— Internal Consumption ——— Exports

Source: Tea Statistics of India (Tea Board, various issues).

been possible partly because of the increase in the number of smaller tea gardens, mainly in Tamil Nadu and Assam. But the growth rate in area continues to be high in many states even during the last sub-period under study, that is, 1998–2004 (see Table 2.12).

TABLE 2.12
Growth of Area under Tea in States of India

States	1951–61	1961–71	1971–81	1981–91	1991–2001	1991–2004	1998–2004
Assam	0.42	1.17	1.08	1.40	1.44	1.18	1.29
West Bengal	0.39	0.68	0.60	0.82	0.84	0.87	1.28
Tripura	0.58	0.74	−0.21	1.01	2.01	2.63	4.48
Bihar	−10.64	−1.48	0.00	−26.22	51.97	41.47	17.45
Uttar Pradesh	−2.21	−1.36	−0.08	−6.69	1.69	3.69	5.18
Himachal Pradesh	−0.31	0.00	0.00	−6.82	1.15	1.00	0.16
Manipur				6.05	12.07	18.57	16.19
Sikkim				−1.15	5.78	1.02	−0.59
Arunachal Pradesh				59.29	7.33	9.59	−2.01
Nagaland					32.41	35.96	26.10
Odisha					0.00	0.00	0.00
Meghalaya							9.65
Mizoram							13.54
North India	0.33	0.96	0.89	1.11	1.36	1.26	1.49
Tamil Nadu	−0.20	0.57	0.68	0.41	6.95	5.34	3.02
Kerala	1.82	−0.65	−0.46	−0.26	0.63	0.52	0.16
Karnataka	0.72	0.42	0.16	0.39	0.74	0.57	0.18
South India	0.85	−0.07	0.10	0.10	4.30	3.33	1.99
All India	0.45	0.74	0.74	0.92	3.44	1.67	1.60

Note: Growth rates are compounded.
Source: Calculated from data provided in *Tea Statistics* (Tea Board, various years).

In terms of trends in growth of production of tea (see Table 2.13), for both Assam and West Bengal the 1970s was the best period among the five post independence decades under study. While growth rates of production have been considerably low in almost all the major tea producing states, for Assam it was the worst. Among the southern states, growth in tea production was negative in Kerala and was exceptionally low in Tamil Nadu. The important aspect of the growth performance of the states in recent years is the remarkably poor performance in production in Assam, as well as north India as a whole during 1998–2004.

The 'crisis' in the tea industry, or to be more precise, the productivity-related aspect of the crisis, comes out in bold relief

TABLE 2.13
Growth of Production of Tea in States of India

States	1951–61	1961–71	1971–81	1981–91	1991–2001	1991–2004	1998–2004
Assam	1.94	2.07	3.15	2.76	1.25	0.65	–1.15
West Bengal	0.99	1.90	2.11	1.93	1.87	2.52	1.71
Tripura	3.68	0.96	0.84	4.82	2.36	2.57	2.45
Bihar	–25.87	–2.89	–31.02	58.49	25.96	19.59	39.64
Uttar Pradesh	–1.13	–1.52	–2.39	–0.41	–4.53	–5.60	–5.66
Himachal Pradesh	0.55	–2.87	–3.73	6.91	–1.46	–4.90	–15.66
Manipur				0.00	12.18	9.96	6.36
Sikkim					1.48	3.58	4.99
Arunachal Pradesh					31.84	31.05	14.89
Nagaland					43.68	41.95	36.79
Odisha					26.51	19.47	1.20
Meghalaya							–4.07
Mizoram							20.95
North India	1.59	1.97	2.79	2.55	1.46	1.26	–0.21
Tamil Nadu	4.66	3.76	2.21	5.29	0.99	2.39	3.57
Kerala	4.37	0.78	1.09	3.44	–0.25	–0.55	–0.98
Karnataka	7.37	4.92	1.71	2.87	2.09	1.68	0.48
South India	4.57	2.43	1.75	4.54	0.60	1.46	2.12
All India	2.19	2.08	2.55	3.01	1.25	1.31	0.36

Note: Growth rates are compounded.
Source: Calculated from data provided in *Tea Statistics* (Tea Board, various years).

when the growth rate of yield of tea is compared across various states and group of states for different decades under consideration (Table 2.14). For the states of Assam and West Bengal, the 1970s was the decade of highest growth in yield per hectare. At the all India level, however, growth of yield was higher in the 1990s than the previous decade because of the high growth of yield in the southern states. Again, the growth rate of yield has been observed to be negative at the all India level as well as in north India, while in south India it has registered a very low growth during 1998–2004. Of the 16 tea producing states, growth in yield was negative in as many as seven states during this period, while in major tea producing states like West Bengal, Tamil Nadu and Karnataka it was abysmally low. To put the record straight, the crisis in the tea industry is primarily a crisis of falling productivity more than anything else.

One aspect of the declining productivity and growth in tea plantation is the inadequate replanting of bushes. The desired annual rate of replanting is 2 per cent. However, the present annual rate

TABLE 2.14
Growth of Yield of Tea in States of India

States	1951–61	1961–71	1971–81	1981–91	1991–2001	1991–2004	1998–2004
Assam	1.52	0.89	2.05	1.34	–0.19	–0.53	–2.41
West Bengal	0.60	1.21	1.50	1.10	1.02	1.64	0.42
Tripura	3.10	0.22	1.05	3.76	0.34	–0.06	–1.95
Bihar	–17.05	–1.45	–31.58	116.60	–17.12	–15.47	18.89
Uttar Pradesh	1.10	–0.16	–2.34	6.74	–6.12	–8.95	–10.31
Himachal Pradesh	0.85	–2.88	–3.73	14.75	–2.58	–5.84	–15.79
Manipur				–5.71	0.10	–7.26	–8.47
Sikkim					–4.07	2.53	5.61
Arunachal Pradesh					22.84	19.59	17.24
Nagaland					8.51	4.41	8.48
Odisha					26.51	19.47	1.20
Meghalaya							–12.51
Mizoram							6.52
North India	1.26	1.00	1.88	1.42	0.10	0.00	–1.68
Tamil Nadu	4.86	3.18	1.52	4.86	–5.57	–2.80	0.53
Kerala	2.51	1.43	1.56	3.71	–0.88	–1.07	–1.14
Karnataka	6.59	4.49	1.54	2.48	1.33	1.10	0.30
South India	3.68	2.51	1.64	4.44	–3.54	–1.81	0.14
All India	1.73	1.33	1.81	2.07	–2.12	–0.36	–1.22

Note: Growth rates are compounded.
Source: Calculated from data provided in *Tea Statistics* (Tea Board, various years).

of replanting is only 0.3 per cent. Again, the growth of area under replantation has been negative during 1991–2000 at the all India level as well as in north and south India separately. When we consider the area under replantation as a percentage of total area under tea plantation the dismal performance becomes even more glaring (see Table 2.15). In terms of area under extension there has been robust growth during 1991–2000 in north India, probably due to the growth of small tea plantations in Assam.

Growth Performance of Tea in Assam: A District-level Analysis

As mentioned earlier, in terms of expansion of area under tea there has been significant growth in Assam in recent years, so much so that when we consider the growth rates for the post-Independence decades, the 1990s comes out as the most favourable decade for gains in terms of area. For a plantation crop like tea, this is presumably a

TABLE 2.15
Growth of Area under Extension, Replacement and Replantation

	Growth Rate in the Period				
	1961– 1970	*1971– 1980*	*1981– 1990*	*1991– 2000*	*1961– 2001*
Area under Extension (North)	3.66	5.83	2.68	4.41	0.79
Area Under Replacement (North)	–0.82	10.05	–2.71	13.08	1.93
Area Under Replantation (North)	–13.16	0.12	0.77	–0.03	0.12
Area under Extension (South)	8.74	0.09	5.57	–9.97	–0.81
Area Under Replacement (South)	–4.62	–1.99	26.85	–10.09	5.97
Area Under Replantation (South)	12.60	–11.25	12.46	–30.11	–0.89
Area under Extension (All India)	4.03	5.20	2.81	3.14	0.84
Area Under Replacement (All India)	–0.99	9.59	–1.73	12.81	2.09
Area Under Replantation (All India)	–12.22	–0.28	1.39	–1.14	0.13
Area under Tea Plantation (All India)	0.80	0.63	0.96	2.07	0.90
Percentage of Area under Extension to Total Area under Tea Plantation (All India)	3.19	4.54	1.83	1.05	–0.06
Percentage of Area under Replacement to Total Area under Tea Plantation (All India)	–1.77	8.90	–2.67	10.53	1.18
Percentage of Area under Replantation to Total Area under Tea Plantation (All India)	–12.92	–0.91	0.42	–3.14	–0.77
Percentage of Area under Extension, Replacement and Replantation to Total Area under Tea Plantation (All India)	–1.73	3.21	0.84	0.96	–0.21

Note: Growth rates are compounded.
Source: Calculated from data provided in *Tea Statistics* (Tea Board, various years).

significant positive development in view of declining forest cover, declining land-man ratio as well as competition for land among alternative crops and alternative industries. However, the regional and spatial concentration of this expansion of area needs careful investigation, as much of the new land under cultivation has come through the small tea plantations. The district-level analysis of growth in area, production and yield of tea has been restricted to the period 1980–2004, for reasons of data availability. It is clear that the 1990s have been a favourable decade for expansion of area under tea, both in the north as well as in the south, but this expansion was highly concentrated in few pockets. For example, in the 1980s, growth of area in Assam was much higher in Goalpara and Sivasagar districts, while the high growth of the 1990s was confined to Dibrugarh and Sivasagar in Assam and the Dooars in West Bengal. Relatively higher growth rates of area under tea was also noticed in the Nilgiris in Tamil Nadu, Palghat in Kerala and Coorg in Karnataka, while during the same period area under tea stagnated or declined in many other districts within the same states. In Assam, the high growth rates of area under tea in the districts closely follow the growth pattern of small tea gardens. Even if the last sub-period under study, that is, 1998–2004 is considered, the high growth in area is noticed in Goalpara, Dibrugarh and Sivasagar (Table 2.16).

TABLE 2.16
Growth Rate of Area under Tea in Tea Producing Districts

State/Districts	1980–2004	1981–90	1991–2000	1991–2004	1998–2004
Darrang	0.62	1.32	0.31	0.25	0.24
Goalpara	0.96	3.98	0.57	1.08	1.66
Kamrup	0.16	1.36	–0.81	0.01	0.90
Dibrugarh	1.97	1.35	3.37	3.28	1.88
Nowgong	0.48	1.32	0.19	0.36	0.65
Sivasagar	1.55	2.00	1.32	1.64	1.18
Cachar	–0.33	0.86	–1.51	–0.45	0.71
Assam	1.16	1.39	1.44	1.65	1.22
Darjeeling	–0.64	0.43	–1.66	–1.06	–0.15
Dooars (a)	2.94	1.75	5.06	5.24	4.49
Terai (b)	0.52	0.69	0.43	0.60	0.79
West Bengal	0.78	0.77	0.77	1.28	1.55
Others Total	1.98	–0.48	14.75	10.32	4.94
North India	1.06	1.14	1.42	1.63	1.41
Coimbatore	0.88	0.18	1.79	1.18	0.19
Kanyakumari	0.49	1.98	0.00	–0.01	–0.04
Madurai	0.14	–0.02	0.12	0.28	0.59
Nilgiris	4.68	0.73	10.13	8.32	3.26
Tirunelveli	0.03	0.00	0.00	0.00	0.00
Tamil Nadu	3.66	0.56	7.85	6.49	2.65
Wynaad	0.13	–0.01	0.36	0.09	–0.40
Idukki	0.70	–0.19	1.64	0.99	0.25
Kottayam	–5.37	–0.85	–10.41	–5.74	0.00
Palghat	1.41	0.37	2.06	1.30	0.25
Quilon	–0.73	–1.81	–0.68	–0.43	0.00
Trichur	0.97	0.34	1.63	1.04	0.11
Trivandrum	–0.49	–0.55	–0.74	–0.42	0.00
Kerala	0.29	–0.29	0.82	0.51	0.13
Coorg	2.65	2.05	1.50	0.98	0.33
Chikmagalur	0.47	0.40	0.72	0.48	0.15
Hassan	–0.11	–0.37	0.00	0.00	0.00
Karnataka	0.61	0.42	0.69	0.45	0.15
South India	2.22	0.16	4.83	4.01	1.75
All India	1.29	0.95	2.09	2.10	1.48

Note: i) Growth rates are compounded.

ii) Major Tea producing States and Regions are italicised (Tables 2.16–2.36)

Source: Calculated from data provided in *Tea Statistics* (Tea Board, various years).

The trends in growth of production of tea reveal the stagnation of production performance at the district level (Table 2.17). For the entire period under study, that is, 1980–2004, growth rates were higher in the southern states in general and Tamil Nadu in particular, in comparison to that of Assam and West Bengal. The districts that

TABLE 2.17
Growth Rate of Production of Tea in Tea Producing Districts

State/Districts	1980–2004	1981–90	1991–2000	1991–2004	1998–2004
Darrang	1.23	3.58	−0.34	−0.30	−0.49
Goalpara	1.89	3.74	0.62	0.36	−0.65
Kamrup	0.50	2.79	−0.97	−0.77	−1.87
Dibrugarh	1.69	2.03	1.28	1.62	1.49
Nowgong	1.44	2.77	0.40	−0.78	−3.51
Sivasagar	1.92	3.50	2.76	1.20	−2.63
Cachar	2.06	2.73	2.52	0.72	−3.75
Assam	1.79	2.89	1.47	0.95	−0.75
Darjeeling	−1.58	−0.55	−5.35	−3.01	0.74
Dooars (a)	5.55	3.46	8.17	9.33	10.29
Terai (b)	1.04	2.46	1.33	0.83	−0.58
West Bengal	1.73	1.33	2.08	2.33	2.15
Others Total	5.17	6.37	3.72	3.74	4.20
North India	1.82	2.47	1.67	1.40	0.20
Coimbatore	1.46	2.64	2.34	0.96	−2.06
Kanyakumari	−2.00	3.84	−2.32	−3.03	−6.33
Madurai	−0.03	5.24	−0.43	−1.24	−3.41
Nilgiris	5.30	9.18	2.45	3.56	6.13
Tirunelveli	0.12	7.66	−8.22	−4.61	1.37
Tamil Nadu	3.94	6.88	2.19	2.78	4.10
Wynaad	2.26	2.45	2.15	1.95	−0.01
Idukki	1.25	3.60	1.12	−0.26	−2.25
Kottayam	8.57	16.57	1.86	8.74	15.04
Palghat	2.86	4.83	4.50	1.93	−1.67
Quilon	−3.19	−6.00	0.45	−1.47	−5.44
Trichur	2.02	6.78	3.20	0.01	−5.61
Trivandrum	−5.43	−1.67	−2.65	−7.95	−10.52
Kerala	1.40	3.32	1.38	0.20	−1.83
Coorg	3.82	2.39	4.78	3.62	1.26
Chikmagalur	1.81	1.91	2.15	2.33	1.54
Hassan	2.52	3.86	3.59	1.07	−3.52
Karnataka	2.20	2.37	2.79	2.23	0.44
South India	2.98	5.38	1.93	1.97	2.22
All India	2.07	3.10	1.72	1.53	0.70

Note: Growth rates are compounded.
Source: Calculated from data provided in *Tea Statistics* (Tea Board, various years).

have experienced relatively better growth in production include Cachar, Sivasagar and Goalpara in Assam, and the Dooars in West Bengal. Moving to growth in various sub-periods, in the 1980s almost all the tea producing districts of Assam registered a growth of more than 2 per cent per annum. Growth rates of tea production were particularly high in Goalpara, Sivasagar and Darrang. During 1991–2000, however, not only did growth of production at the state level slow down considerably, of the seven districts under study, growth was negative in two; while in two others it was negligible. When growth in production during 1991–2004 is taken into account the dismal performance on the tea production front becomes all the more visible in all the districts of Assam, except Dibrugarh (including Lakhimpur). More importantly, growth performance has further worsened during 1998–2004. In as many as six of the seven districts, as well as for Assam as a whole, growth rate of production of tea has been negative (Table 2.17).

This poor performance also gets reflected in the district level estimates of growth in yield per hectare. While yield improvement has not been at the desired level since 1980s itself, the relative performance of the districts of Assam was much better in the 1980s than in the 1990s (Table 2.18).

During 1991–2000, growth of yield slumped in as many as three of the seven districts in Assam. If a longer period of 1991–2004 is taken into consideration negative growth was recorded in six districts, and in the latest sub-period of 1998–2004 growth was negative in all the districts of Assam. Growth rate of yield per hectare has declined considerably during this period in West Bengal, and apart from Tamil Nadu, other southern states have also suffered heavily on the productivity front. The analysis of growth performance of the tea sector clearly brings out the fact that declining productivity in the sector at the all India level gets manifested in the tea gardens of Assam much more pronouncedly than in any other state or region. Within Assam, while there have been some gains in terms of area in few districts, the dismal performance in productivity growth has been all pervasive, particularly in the last few years.

Size-Class-wise Distribution of Tea Gardens

The temporal and spatial variations in the growth performance of the tea sector, all India and within Assam, brought out clearly by

TABLE 2.18
Growth Rate of Yield of Tea in Tea Producing Districts

State/Districts	1980–2004	1981–90	1991–2000	1991–2004	1998–2004
Darrang	0.60	2.23	−0.65	−0.55	−0.73
Goalpara	0.91	−0.22	0.05	−0.70	−2.27
Kamrup	0.35	1.40	−0.16	−0.78	−2.75
Dibrugarh	−0.27	0.66	−2.02	−1.62	−0.38
Nowgong	0.95	1.43	0.21	−1.14	−4.13
Sivasagar	0.35	1.46	1.42	−0.44	−3.77
Cachar	2.39	1.85	4.09	1.18	−4.43
Assam	0.61	1.48	0.03	−0.69	−1.95
Darjeeling	−0.95	−0.98	−3.76	−1.97	0.89
Dooars (a)	2.52	1.68	2.95	3.88	5.56
Terai (b)	0.51	1.76	0.89	0.22	−1.36
West Bengal	0.94	0.55	1.30	1.03	0.59
Others Total	3.12	6.88	−9.62	−5.96	−0.71
North India	0.75	1.32	0.25	−0.23	−1.19
Coimbatore	0.57	2.46	0.54	−0.21	−2.24
Kanyakumari	−2.48	1.83	−2.32	−3.01	−6.29
Madurai	−0.17	5.26	−0.55	−1.52	−3.97
Nilgiris	0.59	8.39	−6.97	−4.38	2.78
Tirunelveli	0.09	7.66	−8.22	−4.61	1.37
Tamil Nadu	0.26	6.28	−5.26	−3.49	1.41
Wynaad	2.12	2.46	1.80	1.86	0.39
Idukki	0.54	3.81	−0.52	−1.24	−2.50
Kottayam	14.73	17.56	13.69	15.37	15.04
Palghat	1.44	4.45	2.38	0.62	−1.92
Quilon	−2.48	−4.28	1.14	−1.04	−5.44
Trichur	1.04	6.42	1.53	−1.01	−5.70
Trivandrum	−4.96	−1.12	−1.93	−7.56	−10.52
Kerala	1.11	3.62	0.56	−0.31	−1.96
Coorg	1.15	0.33	3.23	2.61	0.93
Chikmagalur	1.33	1.50	1.42	1.84	1.39
Hassan	2.64	4.25	3.59	1.07	−3.52
Karnataka	1.58	1.94	2.08	1.78	0.29
South India	0.75	5.21	−2.77	−1.96	0.47
All India	0.77	2.12	−0.36	−0.56	−0.78

Note: Growth rates are compound growth rates.
Source: Calculated from data provided in *Tea Statistics* (Tea Board, various years).

the preceding analysis is associated with the changing structure of production in tea plantations. An important aspect of this change is the growth in the number of small tea gardens in some of the districts. To understand this dimension of the reorganisation of the tea industry, at times linked with the global process of outsourcing and flexibilisation, the changing size-class-wise distribution of tea gardens at the district, state, regional as well as all India level have been presented in Tables 2.19 through 2.22. The distribution of the number of tea gardens, area under tea and production of tea across different size-classes of tea gardens have been looked at four different years, viz., 1980, 1991, 2000 and 2003, the latest year for which data is available.

In 1980, at the all India level, around 90 per cent of the tea gardens were of less than 50 ha size, but the numerical preponderance of smaller tea gardens was largely confined to south India. Around 45 per cent of area was under the largest category of tea gardens, that is, above 400 ha, and these gardens contributed nearly 51 per cent of total production of tea. The tea gardens in the size-class of 200 to 400 ha occupied 35 per cent of area and their share in total production was also 35 per cent. At the state level too, the largest two categories of tea gardens controlled 65 per cent of area in south India and contributed nearly 80 per cent of the production. In West Bengal these two classes of gardens maintained their dominance in all the three aspects — number, size and production. In Assam the most numerous category was those operating between 200 to 400 ha. Tea gardens larger than 400 ha had a 52 per cent share in area and 58 per cent share in production. At the district level, the significance of this category was more in districts like Darrang and Lakhimpur, while in some others, such as Dibrugarh, Nagaon and Cachar, the largest category was the size-class 200–400 ha. In Goalpara, where the share of very small gardens was already sizeable, the majority of tea gardens were between 100 and 200 ha in size (Table 2.19). Thus, even in the early 1980s, the size-class-wise distribution of gardens in the districts of Assam was much varied and differentiated, a feature that remained by and large unchanged even in the early 1990s (Table 2.20). In 1991, for example, in Lakhimpur and Darrang, around 50 per cent of the gardens were of more than 400 ha in size, while some other gardens of 50–100 ha accounted for 14–27 per cent of total gardens. In terms of contribution to total production, however, in all the districts except Goalpara and Sivasagar, the largest sized gardens contributed more than 60 per cent.

The structure of production considerably altered after the inclusion of small gardens in the tea statistics. At the all India level, the share of small tea estates, defined as those operating on land up to 10.12 ha, was 99 per cent in 2000. The picture in Assam changes as well, with the latter category accounting for nearly 98 per cent of all gardens. In the districts, small tea gardens became numerically preponderant, but their share in area as well as in production remained marginal except for Goalpara, Dibrugarh, Sivasagar and Karbi Anglong. While the share of the largest sized garden reduced to less than 1 per cent in number, these gardens continued to contribute 58 per cent of produce and controlled nearly 54 per cent of area in Assam. At the district level, as noted earlier, there was considerable variation in the significance of the three largest categories in terms of their contribution in production and share in area (Table 2.21). The picture remained more or less same in 2003, at least in Assam (Table 2.22).

Size-Productivity Relationship in Tea Gardens

Although, in the case of annual crops, a negative relationship between farm-size and productivity is noticed by some studies,[3] in the case of the tea sector, by and large, the relationship between size and productivity is considered to be positive. Misra (1986), in a study based on the Terai-Dooars region in West Bengal, reported a positive relationship between size and productivity. His result was in conformity with some earlier studies and observations.[4] However, he qualifies the finding by stating that (*a*) yield is influenced by other factors apart from size and (*b*) 'by the very nature of historical development of plantations, the best areas were planted first in large area and cultivation was later extended to less productive areas' (Misra 1986: 43).

Although a detailed investigation into the size-productivity relationship is beyond the scope of the present exercise, the average yield per hectare of area of different size-class of tea gardens has been presented in Tables 2.23 through 2.27. The purpose of the

---

[3] For a recent review of the literature, see Dyer (1996).
[4] *The Report of the Committee on Tea Marketing* (appointed by the Government of India), 1980; *The Report of the Official Team on the Tea Industry*, New Delhi, 1952; *Report of the Special Officer for Cachar and Tripura on Tea*, 1954; *Report of the Plantation Inquiry Commission*, New Delhi, 1956 (all cited in Misra 1986) supported the view that larger tea estates are more productive.

TABLE 2.19
Size-Class-wise Distribution of Number, Area and Production of Tea: 1980

District/ States	Up to 8.09 Hectare			Above 8.09 to 50 Hectare			Above 50 to 100 Hectare			Above 100 to 200 Hectare			Above 200 to 400 Hectare			Above 400 Hectare		
	No.	Area	Prod.	No.	Area	Prod.	No.	Area	Prod.	No.	Area	Prod.	No.	Area	Prod.	No.	Area	Prod.
Darrang	1.08	0.02	0.00	3.23	0.23	0.01	4.30	0.91	0.40	15.05	5.93	4.12	35.48	30.52	29.44	40.86	62.40	66.02
Goalpara	23.08	0.16	0.00	0.00	0.00	0.00	0.00	0.00	0.00	38.46	31.98	20.15	30.77	48.67	60.26	7.69	19.19	19.59
Kamrup	0.00	0.00	0.00	26.67	3.43	0.11	6.67	1.99	0.38	26.67	19.54	5.60	26.67	31.93	27.90	13.33	43.11	66.01
Lakhimpur	0.00	0.00	0.00	23.08	2.20	0.94	7.69	1.84	1.13	7.69	3.98	7.02	23.08	24.63	27.71	38.46	67.35	63.20
Dibrugarh	6.75	0.11	0.06	12.66	1.18	0.93	10.55	3.18	2.84	20.25	11.36	10.44	27.85	31.58	30.30	21.94	52.57	55.43
Nowgong	0.00	0.00	0.00	13.04	1.73	0.63	17.39	4.38	3.09	8.70	4.51	3.38	34.78	34.34	33.33	26.09	55.04	59.58
Sivasagar	4.48	0.12	0.03	17.54	2.47	3.51	19.40	6.72	5.95	18.66	11.33	10.25	22.01	30.85	30.89	17.91	48.51	49.36
Cachar	0.87	0.01	0.00	7.83	0.90	0.35	10.43	2.12	0.81	20.87	10.83	8.31	39.13	40.48	38.84	20.87	45.66	51.69
North Cachar	0.00	0.00	0.00	0.00	0.00	0.00	0.00	0.00	0.00	0.00	0.00	0.00	0.00	0.00	0.00	0.00	0.00	0.00
KAnglong	0.00	0.00	0.00	0.00	0.00	0.00	0.00	0.00	0.00	0.00	0.00	0.00	0.00	0.00	0.00	0.00	0.00	0.00
Assam	4.25	0.08	0.03	12.74	1.40	1.27	12.74	3.60	2.77	19.05	10.33	8.62	28.57	32.80	31.55	22.65	51.79	55.77
Darjeeling	9.71	0.15	0.03	5.83	1.04	0.09	8.74	3.72	2.76	34.95	28.34	28.17	33.98	50.13	51.74	6.80	16.63	17.22
Dooars (a)	1.30	0.02	0.01	1.95	0.12	0.03	3.25	0.64	0.26	10.39	4.25	3.18	35.71	28.37	27.02	47.40	66.59	69.50
Terai (b)	2.08	0.04	0.01	6.25	0.65	0.17	6.25	2.23	2.60	37.50	24.83	27.72	35.42	47.60	42.64	12.50	24.64	26.86
West Bengal	4.26	0.05	0.01	3.93	0.38	0.05	5.57	1.47	0.78	22.95	11.73	8.48	35.08	35.21	31.24	28.20	51.16	59.44
Tripura	1.75	0.07	0.00	15.79	4.55	1.68	31.58	20.50	13.02	42.11	53.96	68.11	8.77	20.92	17.19	0.00	0.00	0.00
Bihar	0.00	0.00	0.00	33.33	8.28	5.00	0.00	0.00	0.00	33.33	29.19	0.00	33.33	62.53	95.00	0.00	0.00	0.00

Uttar Pradesh	22.58	2.22	0.00	41.94	15.30	0.00	19.35	27.66	0.00	9.68	26.22	100.00	6.45	28.60	0.00	0.00	0.00	0.00
Manipur	0.00	0.00	0.00	0.00	0.00	0.00	100.00	100.00	0.00	0.00	0.00	0.00	0.00	0.00	0.00	0.00	0.00	0.00
Sikkim	0.00	0.00	0.00	0.00	0.00	0.00	0.00	0.00	0.00	100.00	0.00	0.00	0.00	0.00	0.00	0.00	0.00	0.00
Arunachal Pr.	100.00	100.00	0.00	0.00	0.00	0.00	0.00	0.00	0.00	0.00	100.00	0.00	0.00	0.00	0.00	0.00	0.00	0.00
Nagaland	0.00	0.00	0.00	0.00	0.00	0.00	0.00	0.00	0.00	0.00	0.00	0.00	0.00	0.00	0.00	0.00	0.00	0.00
Odisha	0.00	0.00	0.00	0.00	0.00	0.00	0.00	0.00	0.00	0.00	0.00	0.00	0.00	0.00	0.00	0.00	0.00	0.00
Himachal Pr.	0.00	0.00	0.00	0.00	0.00	0.00	0.00	0.00	0.00	0.00	0.00	0.00	0.00	0.00	0.00	0.00	0.00	0.00
Mizoram	0.00	0.00	0.00	0.00	0.00	0.00	0.00	0.00	0.00	0.00	0.00	0.00	0.00	0.00	0.00	0.00	0.00	0.00
Meghalaya	0.00	0.00	0.00	0.00	0.00	0.00	0.00	0.00	0.00	0.00	0.00	0.00	0.00	0.00	0.00	0.00	0.00	0.00
Others Total	9.57	0.57	0.00	24.47	6.86	1.57	26.60	21.20	11.94	30.85	47.27	70.24	8.51	24.10	16.25	0.00	0.00	0.00
North India	4.68	0.08	0.02	11.39	1.24	0.90	11.99	3.44	2.25	21.00	11.82	9.14	28.66	33.30	31.31	22.28	50.12	56.37
Tamil Nadu	96.05	19.76	0.00	2.33	7.45	3.53	0.30	4.07	3.61	0.40	11.00	13.66	0.69	36.87	51.25	0.24	20.85	27.95
Kerala	95.18	6.04	0.00	1.82	4.15	1.25	0.46	3.81	2.01	0.71	11.88	10.83	1.36	44.86	51.83	0.46	29.25	34.07
Karnataka	13.33	0.48	0.00	20.00	3.39	0.67	13.33	7.90	4.92	33.33	43.24	49.94	20.00	44.99	44.49	0.00	0.00	0.00
South India	95.60	12.67	0.00	2.16	5.76	2.40	0.38	4.04	2.92	0.56	12.24	13.57	0.97	40.92	51.29	0.32	24.37	29.82
All India	86.70	2.58	0.02	3.07	2.14	1.20	1.52	2.56	2.38	2.57	11.90	10.02	3.68	34.81	35.28	2.47	45.01	51.11

Source: Tea Statistics (various years).

TABLE 2.20
Size-Class-wise Distribution of Number, Area and Production of Tea: 1991

District/States	Up to 8.09 Hectare			Above 8.09 to 50 Hectare			Above 50 to 100 Hectare			Above 100 to 200 Hectare			Above 200 to 400 Hectare			Above 400 Hectare		
	No.	Area	Prod.	No.	Area	Prod.	No.	Area	Prod.	No.	Area	Prod.	No.	Area	Prod.	No.	Area	Prod.
Darrang	0.00	0.00	0.00	0.00	0.00	0.00	3.33	0.57	0.30	16.67	5.82	4.88	28.89	21.63	22.60	51.11	71.98	72.22
Goalpara	0.00	0.00	0.00	0.00	0.00	0.00	0.00	0.00	0.00	30.00	15.70	12.81	60.00	65.55	65.32	10.00	18.74	21.87
Kamrup	0.00	0.00	0.00	0.00	0.00	0.00	8.33	2.86	0.77	41.67	19.53	1.69	33.33	34.09	34.27	16.67	43.52	63.27
Lakhimpur	0.00	0.00	0.00	8.33	0.44	0.00	16.67	3.80	2.04	8.33	3.49	4.03	16.67	14.57	16.10	50.00	77.71	77.83
Dibrugarh	3.64	0.05	0.06	11.36	1.50	1.38	13.18	3.46	5.14	20.45	10.74	10.38	23.64	23.83	23.50	27.73	60.41	59.52
Nowgong	0.00	0.00	0.00	4.35	0.32	0.00	26.09	6.24	5.47	4.35	2.00	1.12	34.78	31.67	29.00	30.43	59.77	64.42
Sivasagar	1.20	0.06	0.02	14.06	1.77	1.17	19.28	6.70	6.80	18.07	10.12	14.70	23.29	25.69	25.90	24.10	55.66	51.41
Cachar	0.00	0.00	0.00	8.60	1.03	0.12	6.45	1.52	0.81	17.20	6.93	3.18	40.86	39.73	35.66	26.88	50.79	60.23
North Cachar	0.00	0.00	0.00	0.00	0.00	0.00	0.00	0.00	0.00	0.00	0.00	0.00	0.00	0.00	0.00	0.00	0.00	0.00
KAnglong	0.00	0.00	0.00	0.00	0.00	0.00	0.00	0.00	0.00	0.00	0.00	0.00	0.00	0.00	0.00	0.00	0.00	0.00
Assam	1.55	0.03	0.03	9.87	1.14	0.82	13.40	3.68	4.01	18.48	8.89	9.25	27.36	27.00	25.76	29.34	59.25	60.13
Darjeeling	0.00	0.00	0.00	4.49	0.70	0.17	12.36	4.89	2.99	38.20	27.29	23.39	37.08	49.79	56.99	7.87	17.32	16.46
Dooars (a)	4.46	0.07	0.03	1.91	0.11	0.09	2.55	0.05	0.38	7.64	2.95	2.06	30.57	21.83	20.39	52.87	74.52	77.06
Terai (b)	16.13	0.26	0.13	6.45	0.42	0.29	6.45	2.10	3.04	24.19	16.61	12.75	30.65	41.93	40.74	16.13	38.67	43.04

West Bengal	5.52	0.08	0.04	3.57	0.27	0.12	6.17	1.57	0.98	19.81	9.45	5.36	32.47	29.89	26.36	32.47	58.74	67.13
Tripura	0.00	0.00	0.00	13.46	3.57	6.78	40.38	25.95	28.92	36.54	47.33	38.54	7.69	15.92	15.24	1.92	7.22	10.51
Bihar	0.00	0.00	0.00	100.00	100.00	100.00	0.00	0.00	0.00	0.00	0.00	0.00	0.00	0.00	0.00	0.00	0.00	0.00
Uttar Pradesh	0.00	0.00	0.00	0.00	0.00	0.00	0.00	0.00	0.00	0.00	0.00	0.00	0.00	0.00	0.00	0.00	0.00	0.00
Manipur	0.00	0.00	0.00	50.00	13.19	0.00	1500	0.69	500.00	1250	100.00	100.00	0.00	0.00	0.00	0.00	0.00	0.00
Sikkim	0.00	0.00	0.00	0.00	0.00	0.00	0.00	0.00	0.00	100.00	0.00	0.00	100.00	0.00	0.00	0.00	0.00	0.00
Arunachal Pr.	25.00	2.19	0.00	50.00	24.56	84.31	0.00	0.00	0.00	25.00	29.39	15.69	0.00	0.00	0.00	0.00	0.00	0.00
Nagaland	0.00	0.00	0.00	100.00	100.00	100.00	0.00	0.00	0.00	0.00	0.00	0.00	0.00	100.00	0.00	0.00	0.00	0.00
Odisha	0.00	0.00	0.00	0.00	0.00	0.00	0.00	0.00	0.00	0.00	0.00	0.00	100.00	100.00	100.00	0.00	0.00	0.00
Himachal Pr.	0.00	0.00	0.00	0.00	0.00	0.00	0.00	0.00	0.00	0.00	0.00	0.00	0.00	0.00	0.00	0.00	0.00	0.00
Mizoram	0.00	0.00	0.00	0.00	0.00	0.00	0.00	0.00	0.00	0.00	0.00	0.00	0.00	0.00	0.00	0.00	0.00	0.00
Meghalaya	0.00	0.00	0.00	0.00	0.00	0.00	0.00	0.00	0.00	0.00	0.00	0.00	0.00	0.00	0.00	0.00	0.00	0.00
Others Total	1.61	0.08	0.00	19.35	5.13	9.09	33.87	22.75	27.42	35.48	48.51	38.89	8.06	17.20	14.64	1.61	6.33	9.97
North India	2.69	0.05	0.03	8.62	0.96	0.71	12.51	3.44	3.39	19.93	9.88	8.45	27.71	27.67	25.82	28.64	58.00	61.60
Tamil Nadu	96.30	21.09	0.00	2.04	6.32	2.92	0.31	4.08	3.18	0.38	101.93	11.30	0.75	39.32	54.42	0.22	18.89	28.18
Kerala	95.88	7.32	0.00	1.29	2.91	0.92	0.44	3.77	1.59	0.63	10.96	9.38	1.24	42.23	45.59	0.51	32.81	42.53
Karnataka	9.09	0.51	0.00	40.91	3.34	1.94	13.64	10.68	5.61	18.18	31.73	33.76	18.18	53.74	58.68	0.00	0.00	0.00
South India	95.97	14.21	0.00	1.84	4.67	1.97	0.38	4.11	2.52	0.51	11.17	11.14	0.97	41.04	50.48	0.33	24.81	33.89
All India	87.59	2.75	0.03	2.45	1.67	0.97	1.47	3.57	3.21	2.26	10.13	9.01	3.37	30.22	30.92	2.87	51.67	55.87

Source: Tea Statistics (various years).

TABLE 2.21

Size-Class-wise Distribution of Number, Area and Production of Tea: 2000

District/States	Up to 10.12 Hectare			Above 10.12 to 50 Hectare			Above 50 to 100 Hectare			Above 100 to 200 Hectare			Above 200 to 400 Hectare			Above 400 Hectare		
	No.	Area	Prod.	No.	Area	Prod.	No.	Area	Prod.	No.	Area	Prod.	No.	Area	Prod.	No.	Area	Prod.
Darrang	88.66	1.97	1.29	0.00	0.00	0.00	0.36	0.56	0.55	2.17	6.43	5.91	2.41	16.49	16.78	6.39	74.54	75.48
Goalpara	95.18	9.62	8.80	0.80	1.50	0.00	0.00	0.00	0.00	0.80	8.09	6.34	1.61	25.90	22.22	1.61	54.88	62.65
Kamrup	75.93	1.10	1.02	5.56	2.15	1.30	1.85	2.38	1.16	5.56	15.60	6.25	5.56	26.09	29.61	5.56	52.67	60.65
Lakhimpur	96.01	5.01	1.86	0.31	0.46	0.33	0.61	1.83	1.99	0.00	0.00	0.00	0.61	10.30	9.87	2.45	82.41	85.95
Dibrugarh	98.86	26.37	18.77	0.16	0.96	0.47	0.17	2.99	2.39	0.25	8.63	8.73	0.25	16.13	17.09	0.30	44.92	52.55
Nowgong	89.20	2.84	2.43	0.47	0.26	0.25	2.82	5.84	4.00	0.47	1.84	0.83	3.76	31.14	31.90	3.29	58.08	60.58
Sivasagar	98.37	16.18	16.83	0.34	2.09	1.64	0.24	3.82	3.79	0.32	9.46	8.85	0.32	20.06	21.96	0.40	48.39	46.93
Cachar	49.51	1.06	1.32	4.85	0.91	0.28	4.85	1.97	1.56	6.80	6.53	4.82	18.45	31.69	27.96	15.53	57.84	64.06
North Cachar	0.00	0.00	0.00	0.00	0.00	0.00	0.00	0.00	0.00	0.00	0.00	0.00	50.00	32.49	39.00	50.00	67.51	61.00
KAnglong	90.91	23.22	35.01	2.10	6.26	4.63	4.20	22.20	21.23	2.10	21.03	22.67	0.00	0.00	0.00	0.70	27.29	16.45
Assam	98.01	14.66	12.08	0.28	1.14	0.69	0.26	2.83	2.39	0.37	7.95	7.35	0.47	19.89	20.19	0.61	53.52	57.31
Darjeeling	0.00	0.00	0.00	2.35	0.40	0.13	10.59	4.20	3.94	37.65	27.91	26.35	41.18	48.87	51.24	8.24	18.62	18.33
Dooars (a)	71.01	1.79	4.15	1.10	0.27	0.19	1.10	0.62	0.62	2.57	3.13	2.50	7.52	17.94	16.83	16.70	76.24	75.71
Terai (b)	92.86	23.18	28.96	0.88	0.63	0.23	0.66	2.31	1.68	1.65	10.30	8.37	2.53	30.77	25.34	1.43	32.81	35.42
West Bengal	80.00	5.60	9.85	1.04	0.36	0.20	1.36	1.52	1.04	3.96	8.47	5.12	6.43	25.35	20.62	7.21	58.70	63.17

Tripura	81.51	7.70	2.94	4.11	5.01	3.73	5.14	16.91	16.56	6.85	38.58	39.99	2.05	24.11	22.94	0.34	7.69	13.84
Bihar	99.59	98.00	79.55	0.41	2.00	20.45	0.00	0.00	0.00	0.00	0.00	0.00	0.00	0.00	0.00	0.00	0.00	0.00
Uttar Pradesh	18.18	1.22	0.00	36.36	4.87	4.55	9.09	7.87	6.44	18.18	36.99	24.62	18.18	49.06	64.39	0.00	0.00	0.00
Manipur	84.62	36.38	0.00	5.13	4.52	0.00	5.13	21.06	0.00	0.00	0.00	0.00	5.13	38.04	87.50	0.00	0.00	0.00
Sikkim	98.65	41.55	0.00	0.00	0.00	0.00	0.00	0.00	0.00	1.35	58.45	100.00	0.00	0.00	0.00	0.00	0.00	0.00
Arunachal Pr.	46.00	9.97	0.00	32.00	24.82	17.32	12.00	25.02	37.36	6.00	16.18	11.78	4.00	26.01	33.53	0.00	0.00	0.00
Nagaland	92.55	57.66	0.00	4.26	12.93	16.28	0.00	0.00	0.00	1.06	6.51	46.51	2.13	22.90	37.21	0.00	0.00	0.00
Odisha	0.00	0.00	0.00	0.00	0.00	0.00	0.00	0.00	0.00	0.00	0.00	0.00	100.00	100.00	100.00	0.00	0.00	0.00
Himachal Pr.	99.95	96.82	95.27	0.05	3.18	0.00	0.00	0.00	0.00	0.00	0.00	0.00	0.00	0.00	0.00	0.00	0.00	0.00
Mizoram	41.67	8.44	0.00	50.00	71.87	4.73	8.33	19.69	23.08	0.00	0.00	0.00	0.00	0.00	0.00	0.00	0.00	0.00
Meghalaya	53.33	48.15	0.00	46.67	51.85	76.92	0.00	0.00	0.00	0.00	0.00	0.00	0.00	0.00	0.00	0.00	0.00	0.00
Others Toal	97.30	33.51	18.17	1.20	9.97	7.70	0.55	1.66	14.62	0.60	21.01	28.79	0.33	20.83	21.83	0.02	3.01	8.90
North India	97.32	12.98	11.54	0.39	1.31	0.66	0.33	2.85	2.20	0.52	8.66	7.05	0.66	21.43	20.34	0.78	52.76	58.21
Tamil Nadu	99.62	56.39	58.08	0.18	2.94	1.26	0.04	2.59	1.07	0.04	4.73	4.88	0.09	21.87	19.17	0.03	11.49	15.55
Kerala	97.50	13.02	5.29	0.65	2.52	1.61	0.26	3.25	1.35	0.42	10.30	10.22	0.83	39.76	38.89	0.34	31.15	42.64
Karnataka	43.24	3.91	1.15	27.03	4.81	2.26	8.11	9.94	7.53	10.81	30.11	31.51	10.81	51.23	57.56	0.00	0.00	0.00
South India	99.40	41.29	38.93	0.24	2.84	1.40	0.07	2.94	1.34	0.08	7.02	7.36	0.16	28.25	26.77	0.06	17.67	24.20
All India	98.56	19.35	18.21	0.30	1.65	0.84	0.17	2.87	1.99	0.26	8.29	7.13	0.36	22.97	21.90	0.35	44.87	49.93

Source: Tea Statistics (various years).

TABLE 2.22
Size-Class-wise Distribution of Number, Area and Production of Tea: 2003

District/States	Up to 10.12 Hectare			Above 10.12 to 50 Hectare			Above 50 to 100 Hectare			Above 100 to 200 Hectare			Above 200 to 400 Hectare			Above 400 Hectare		
	No.	Area	Prod.	No.	Area	Prod.	No.	Area	Prod.	No.	Area	Prod.	No.	Area	Prod.	No.	Area	Prod.
Darrang	93.90	2.07	1.53	0.06	0.06	0.00	0.19	0.57	0.43	1.17	6.00	5.27	1.23	14.73	14.26	3.44	76.57	78.52
Goalpara	95.44	10.29	9.22	0.76	1.43	0.00	0.00	0.00	0.00	0.76	7.68	5.69	1.52	29.82	26.80	1.52	50.78	58.29
Kamrup	75.93	1.53	2.49	5.56	2.14	0.00	1.85	2.37	0.46	5.56	15.49	6.62	7.41	37.71	35.91	3.70	40.77	54.51
Lakhimpur	97.18	5.21	4.19	0.22	0.45	0.00	0.43	2.03	2.42	0.22	2.81	6.60	0.65	18.08	20.98	1.30	71.42	65.81
Dibrugarh	98.84	27.08	20.54	0.22	1.16	0.62	0.18	3.17	2.69	0.25	8.90	9.25	0.24	16.23	16.52	0.28	43.46	50.38
Nowgong	93.68	3.12	2.88	0.27	0.26	0.00	1.65	5.68	5.65	0.27	2.05	0.98	2.20	30.77	24.57	1.92	58.12	65.92
Sivasagar	98.51	16.61	15.26	0.31	2.00	1.91	0.22	3.80	4.91	0.29	8.90	9.21	0.31	20.49	20.31	0.36	48.21	48.40
Cachar	49.51	1.06	0.85	5.34	1.05	0.12	4.37	1.87	0.31	6.80	5.41	1.77	18.45	32.67	24.03	15.53	57.94	72.92
North Cachar	0.00	0.00	0.00	0.00	0.00	0.00	0.00	0.00	0.00	0.00	0.00	0.00	50.00	32.95	49.53	50.00	67.05	50.47
KAnglong	96.69	28.76	47.49	0.51	3.69	2.66	1.78	23.56	12.43	0.76	20.70	13.76	0.00	0.00	0.00	0.25	23.30	23.67
Assam	98.15	15.19	12.72	0.29	1.19	0.72	0.24	2.90	2.59	0.34	7.75	7.37	0.43	20.22	18.60	0.54	52.75	58.01
Darjeeling	0.00	0.00	0.00	2.35	0.41	0.09	10.59	4.26	3.90	37.65	25.82	28.27	41.18	51.24	50.35	8.24	18.27	17.38
Dooars (a)	94.64	4.12	8.61	0.30	0.29	0.03	0.17	0.54	0.37	0.47	2.76	2.52	1.37	17.20	15.74	3.05	75.10	72.74
Terai (b)	98.83	28.30	35.03	0.12	0.57	0.17	0.07	1.23	1.11	0.28	9.99	12.29	0.48	30.94	29.93	0.21	28.98	21.47
West Bengal	96.43	8.38	16.07	0.21	0.36	0.07	0.21	1.25	0.76	0.71	7.80	6.66	1.18	25.26	21.62	1.26	56.94	54.82

Tripura	96.30	23.66	9.98	1.02	4.85	5.42	0.96	14.71	20.09	1.28	31.12	33.60	0.38	19.51	24.58	0.06	6.16	6.33
Bihar	99.90	98.65	96.37	0.10	1.35	3.63	0.00	0.00	0.00	0.00	0.00	0.00	0.00	0.00	0.00	0.00	0.00	0.00
Uttarakhand	87.50	50.71	0.00	6.25	3.54	0.00	1.25	3.67	6.38	2.50	13.26	13.48	2.50	28.82	80.14	0.00	0.00	0.00
Manipur	98.61	90.98	39.50	1.39	9.02	60.50	0.00	0.00	0.00	0.00	0.00	0.00	0.00	0.00	0.00	0.00	0.00	0.00
Sikkim	75.00	8.72	0.00	0.00	0.00	0.00	0.00	0.00	0.00	25.00	91.28	100.00	0.00	0.00	0.00	0.00	0.00	0.00
Arunachal Pradesh	57.14	8.79	0.00	34.92	28.72	28.08	4.76	16.03	25.33	0.00	0.00	0.00	3.17	46.46	46.59	0.00	0.00	0.00
Nagaland	99.52	94.84	72.31	0.48	5.16	27.69	0.00	0.00	0.00	0.00	0.00	0.00	0.00	0.00	0.00	0.00	0.00	0.00
Odisha	0.00	0.00	0.00	0.00	0.00	0.00	0.00	0.00	0.00	0.00	0.00	0.00	100.00	100.00	100.00	0.00	0.00	0.00
Himachal Pradesh	99.35	70.70	31.40	0.56	18.78	34.71	0.05	6.09	18.51	0.03	4.43	15.37	0.00	0.00	0.00	0.00	0.00	0.00
Mizoram	97.46	78.40	0.00	2.17	11.33	51.28	0.36	10.27	48.72	0.00	0.00	0.00	0.00	0.00	0.00	0.00	0.00	0.00
Meghalaya	84.44	44.44	0.00	15.56	55.56	100.00	0.00	0.00	0.00	0.00	0.00	0.00	0.00	0.00	0.00	0.00	0.00	0.00
Others Total	98.27	50.94	18.03	1.05	8.60	11.28	0.26	8.46	18.04	0.28	15.29	24.07	0.13	14.15	24.37	0.01	2.55	4.21
North India	97.92	15.04	13.86	0.38	1.33	0.73	0.24	2.71	2.33	0.39	8.14	7.48	0.50	21.33	19.65	0.57	51.45	55.95
Tamil Nadu	99.63	57.41	52.84	0.18	2.88	1.09	0.05	2.63	1.63	0.04	4.58	4.24	0.08	19.54	24.78	0.03	12.96	15.42
Kerala	97.50	13.01	4.55	0.65	2.55	1.04	0.26	3.25	0.86	0.42	10.38	8.55	0.83	39.70	36.01	0.34	31.12	49.00
Karnataka	43.75	3.01	2.96	21.88	5.97	1.52	9.38	9.92	8.66	12.50	30.03	32.00	12.50	51.08	54.86	0.00	0.00	0.00
South India	99.42	42.10	39.51	0.23	2.83	1.09	0.07	2.96	1.60	0.08	6.92	5.96	0.15	26.62	28.30	0.06	18.57	23.54
All India	98.71	21.02	20.57	0.30	1.66	0.82	0.15	2.77	2.14	0.22	7.87	7.08	0.31	22.50	21.91	0.30	44.19	47.47

Source: Tea Statistics (various years).

exercise, notwithstanding the limited comparability of the figures in some years because of a change in class-boundaries, is to gain an insight into the changing pattern of changes in average productivity in different categories of gardens at a disaggregated level. The analysis, however, suffers from non-availability of data on production in small tea gardens, particularly for 1980 and 1991. Because of change in the class-boundaries in later years, the productivity of the smallest two size-classes is not comparable across time. For the last sub-period under study, 1998–2003, growth rates of yield have been computed for all the size-classes, including the two smallest classes.

At the all India level, particularly for the years 1999–2003, productivity, defined here, as production per hectare, is highest in

TABLE 2.23
Farm-size-wise Productivity: 1980

District/States	Up to 10.12	10.12– 50.0	50–100	100–200	200–400	Above 400	All
Darrang	n.a.	112.50	794.30	1252.67	1738.58	1906.51	1801.86
Goalpara	n.a.	n.a.	n.a.	984.91	1934.71	1595.39	1562.75
Kamrup	n.a.	46.73	274.19	413.11	1258.78	2205.79	1440.74
Lakhimpur	n.a.	734.18	1060.61	3041.96	1938.98	1617.36	1723.63
Dibrugarh	3812.50	1449.65	1649.97	1701.58	1776.02	1951.88	1851.14
Nowgong	n.a.	479.34	934.85	990.51	1285.24	1433.20	1324.10
Sivasagar	1833.33	1779.02	1109.37	1133.74	1254.08	1274.38	1252.48
Cachar	0.00	381.94	372.43	750.50	937.96	1106.67	977.62
Assam	2515.15	1354.45	1152.99	1250.68	1441.89	1614.51	1499.23
Darjeeling	400.00	54.73	488.83	655.54	680.66	683.03	659.48
Dooars (a)	5000.00	448.72	678.30	1246.16	1586.93	1739.30	1666.46
Terai (b)	1000.00	360.00	1613.28	1544.05	1239.11	1508.31	1383.30
West Bengal	1153.85	206.21	753.82	1029.99	1263.70	1655.00	1424.48
Tripura	n.a.	224.64	386.16	767.96	500.00	n.a.	608.38
Bihar	n.a.	26.32	n.a.	0.00	66.20	n.a.	43.57
Uttarakhand	n.a.	n.a.	n.a.	602.54	n.a.	n.a.	157.98
Sikkim	n.a.	n.a.	n.a.	140.54	n.a.	n.a.	140.54
Others Total	0.00	106.78	263.45	694.81	315.31	n.a.	467.60
North India	1781.82	1050.08	944.67	1119.82	1360.54	1627.28	1446.85
Tamil Nadu	n.a.	709.21	1326.48	1856.93	2079.02	2005.83	1495.92
Kerala	n.a.	412.94	720.61	1244.99	1578.95	1591.94	1366.48
Karnataka	n.a.	375.00	1181.21	2189.95	1875.15	n.a.	1896.13
South India	n.a.	601.48	1044.17	1600.44	1809.48	1766.66	1443.67
All India	9.42	810.03	967.13	1218.04	1465.38	1642.27	1446.22

Source: Calculated from data provided in *Tea Statistics* (Tea Board, various years).

TABLE 2.24
Farm-size-wise Productivity: 1991

District/States	Up to 8.09	8.09–50.0	50–100	100–200	200–400	Above 400	All
Darrang	n.a.	n.a.	1039.82	1674.08	2083.54	2000.74	1994.14
Goalpara	n.a.	n.a.	n.a.	1288.42	1573.88	1843.03	1579.50
Kamrup	n.a.	n.a.	373.63	120.77	1403.14	2028.90	1395.60
Lakhimpur	n.a.	0.00	993.63	2145.83	2049.83	1857.94	1855.00
Dibrugarh	10500.00	1926.88	3116.94	2025.97	2066.91	2065.21	2096.01
Nowgong	n.a.	0.00	1365.80	871.62	1426.68	1679.69	1558.39
Sivasagar	8000.00	1035.78	1592.22	2277.53	1580.34	1447.95	1567.80
Cachar	n.a.	143.32	627.75	544.09	1064.10	1405.64	1185.44
North Cachar	n.a.	n.a.	n.a.	n.a.	n.a.	n.a.	n.a.
Karbi Anglong	n.a.	n.a.	n.a.	n.a.	n.a.	n.a.	n.a.
Assam	9818.18	1257.91	1909.98	1823.24	1672.36	1779.30	1753.12
Darjeeling	n.a.	162.79	403.53	565.05	754.69	626.28	659.38
Dooars (a)	4857.14	1364.86	13303.03	1258.37	1686.16	1866.89	1805.27
Terai (b)	3000.00	1222.22	2563.43	1360.59	1721.64	1971.79	1771.90
West Bengal	3764.71	731.52	989.39	896.21	1393.52	1805.49	1579.89
Tripura	n.a.	1703.88	999.33	730.23	858.54	1304.56	896.74
Bihar	n.a.	4545.45	n.a.	n.a.	n.a.	n.a.	4545.45
Uttarakhand	n.a.	n.a.	n.a.	n.a.	n.a.	n.a.	n.a.
Manipur	n.a.	n.a.	n.a.	n.a.	n.a.	n.a.	173.61
Sikkim	n.a.	n.a.	n.a.	555.56	n.a.	n.a.	555.56
Arunachal Pr.	n.a	767.86	n.a.	119.40	n.a.	n.a.	223.68
Nagaland	n.a.	57.14	n.a.	n.a.	n.a.	n.a.	57.14
Odisha	n.a.	n.a.	n.a.	n.a.	46.73	n.a.	46.73
Himachal Pr.	n.a.	n.a.	n.a.	n.a.	n.a.	n.a.	n.a.
Mizoram	n.a.	n.a.	n.a.	n.a.	n.a.	n.a.	n.a.
Meghalaya	n.a.	n.a.	n.a.	n.a.	n.a.	n.a.	n.a.
Others Total	n.a.	1467.46	999.33	664.48	705.21	1304.56	828.88
North India	5931.03	1236.85	1659.10	1438.58	1569.21	1786.22	1681.85
Tamil Nadu	n.a.	848.83	1431.11	203.43	2540.55	2738.97	1835.74
Kerala	n.a.	588.29	784.24	1593.95	2009.70	2412.78	1861.76
Karnataka	n.a.	1333.33	1203.79	2437.00	2500.94	n.a.	2290.49
South India	n.a.	783.21	1142.26	1855.34	2287.58	2540.24	1859.66
All India	16.34	994.37	1545.44	1526.32	1755.45	1855.34	1715.80

Source: Calculated from data provided in *Tea Statistics* (Tea Board, various years).

the largest size-class of gardens. This clearly indicates the presence of significant economies of scale, particularly for the gardens of more than 400 ha of size. This is true for Assam as well. However, at the district level, some variations in the levels of productivity across size-classes of gardens are noticed. In 1980, in four out of the eight districts, highest yield was recorded in the largest size-class. In 1998,

TABLE 2.25
Farm-size-wise Productivity: 2000

District/States	Up to 10.12 ha	10.12– 50.0 ha	50–100 ha	100–200 ha	200–400 ha	Above 400 ha	All
Darrang	1222.22	n.a.	1818.18	1725.76	1909.14	1900.84	1877.09
Goalpara	1663.66	n.a.	n.a.	1425.00	1561.38	2077.41	1819.94
Kamrup	1157.89	756.76	609.76	500.93	1418.71	1439.05	1249.85
Lakhimpur	701.24	1363.64	2045.45	n.a.	1804.44	1964.21	1883.28
Dibrugarh	1249.89	864.20	1404.53	1775.30	1859.36	2054.34	1755.83
Nowgong	1264.32	1428.57	1010.71	666.67	1510.65	1538.01	1474.61
Sivasagar	1668.73	1257.18	1592.09	1499.86	1755.85	1555.38	1603.83
Cachar	1911.76	477.66	1215.53	1135.47	1356.40	1702.50	1537.30
North Cachar	n.a.	n.a.	n.a.	n.a.	1852.42	1394.38	1543.21
Karbi Anglong	1569.12	769.23	995.18	1122.14	n.a.	627.45	1040.66
Assam	1388.49	1016.47	1424.31	1557.39	1710.63	1804.71	1685.55
Darjeeling	n.a.	173.91	505.52	508.74	564.85	530.40	538.72
Dooars (a)	4274.18	1284.21	1831.80	1478.94	1736.08	1837.41	1850.19
Terai (b)	2631.95	769.23	1530.53	1712.67	1734.78	2274.89	2106.82
West Bengal	2973.56	915.17	1156.15	1021.08	1373.84	1817.71	1689.04
Tripura	370.59	722.89	950.89	1006.65	923.61	1748.53	971.01
Bihar	323.51	4074.07	n.a.	n.a.	n.a.	n.a.	398.52
Uttarakhand	n.a.	230.77	202.38	164.56	324.43	n.a.	247.19
Manipur	36.36	n.a.	n.a.	n.a.	243.48	n.a.	105.84
Sikkim	n.a.	n.a.	n.a.	606.94	n.a.	n.a.	354.73
Arunachal Pr.	n.a.	318.52	740.52	332.39	588.34	n.a.	456.34
Nagaland	n.a.	44.59	n.a.	253.16	57.55	n.a.	35.42
Odisha	n.a.	n.a.	n.a.	n.a.	490.65	n.a.	490.65
Himachal Pr.	527.77	797.30	n.a.	n.a.	n.a.	n.a.	536.34
Mizoram	n.a.	106.76	116.88	n.a.	n.a.	n.a.	99.74
Meghalaya	n.a.	769.23	n.a.	n.a.	n.a.	n.a.	398.86
Others Total	320.52	456.70	741.00	810.07	619.47	1748.53	591.25
North India	1457.02	824.07	1264.23	1334.68	1555.22	1808.55	1639.16
Tamil Nadu	1824.79	758.70	735.45	1825.10	1552.82	2398.24	1771.71
Kerala	758.84	1192.27	775.37	1851.96	1825.49	2555.05	1866.46
Karnataka	746.99	1196.08	1928.91	2666.67	2862.93	n.a.	2548.07
South India	1713.43	898.04	825.29	1905.45	1722.22	2488.23	1817.08
All India	1580.08	852.65	1163.17	1443.37	1601.42	1868.77	1679.18

Source: Calculated from data provided in *Tea Statistics* (Tea Board, various years).

when data on 'unorganised' tea gardens were included, in four out of 11 districts the same conclusion was found to be true, while two others, the second largest size-class, that is, between 200 to 400 ha, exhibited the highest productivity. In 2000, the largest two size-classes recorded the highest productivity in eight of the 11 districts, while in 2003, in six of the districts, the largest size-class was also the class

TABLE 2.26
Farm-size-wise Productivity: 2001

District/States	Up to 10.12 ha	10.12– 50.0 ha	50–100 ha	100–200 ha	200–400 ha	Above 400 ha	All
Darrang	1237.04	n.a.	1512.82	1583.07	2006.23	1911.03	1890.98
Goalpara	1618.62	n.a.	n.a.	1267.86	1551.34	2122.51	1825.99
Kamrup	1184.21	770.27	609.76	696.46	1390.87	1556.72	1335.27
Lakhimpur	730.29	1363.64	1822.92	1716.42	1836.73	2020.07	1905.31
Dibrugarh	1264.72	1107.97	1601.64	1839.62	1813.75	2123.20	1786.90
Nowgong	1312.78	1428.57	1473.80	804.05	1417.72	1540.52	1478.76
Sivasagar	1672.05	1277.16	1743.23	1449.35	1493.94	1588.15	1571.69
Cachar	1867.65	529.41	1498.33	1184.62	1251.91	1725.51	1522.58
North Cachar	n.a.	n.a.	n.a.	n.a.	1609.57	1152.64	1258.30
Karbi Anglong	1564.52	n.a.	858.35	1145.04	n.a.	915.18	1112.70
Assam	1398.64	1108.94	1586.88	1568.85	1601.89	1831.64	1685.23
Darjeeling	n.a.	166.67	576.92	558.44	568.15	566.87	563.86
Dooars (a)	3099.60	1548.12	1538.21	1312.12	1598.20	1770.66	1769.70
Terai (b)	3178.40	769.23	1319.89	1684.55	2007.42	2297.08	2323.27
West Bengal	3152.13	1092.97	980.73	1023.19	1370.15	1769.75	1685.98
Tripura	208.41	1050.38	1007.14	1013.70	1108.33	762.28	903.61
Bihar	669.96	2037.04	n.a.	n.a.	n.a.	n.a.	695.50
Uttarakhand	n.a.	250.00	226.19	174.68	431.30	n.a.	306.18
Manipur	136.48	710.14	n.a.	n.a.	n.a.	n.a.	224.44
Sikkim	n.a.	n.a.	n.a.	635.84	n.a.	n.a.	366.67
Arunachal Pr.	n.a.	950.35	1314.92	n.a.	1067.06	n.a.	981.26
Nagaland	n.a.	961.54	n.a.	n.a.	n.a.	n.a.	129.31
Odisha	n.a.	n.a.	n.a.	n.a.	490.65	n.a.	490.65
Himachal Pr.	404.31	488.64	594.41	625.00	n.a.	n.a.	442.04
Mizoram	0.00	111.11	116.88	n.a.	n.a.	n.a.	102.50
Meghalaya	0.00	525.64	n.a.	n.a.	n.a.	n.a.	398.06
Others Total	356.91	680.89	921.50	878.22	929.63	762.28	685.65
North India	1542.88	965.97	1408.29	1345.37	1503.25	1809.92	1647.14
Tamil Nadu	1822.05	687.18	678.84	1111.36	1841.37	1983.76	1750.76
Kerala	553.85	740.06	626.46	1750.51	1779.09	2421.83	1763.70
Karnataka	734.94	1750.00	2071.09	2730.83	2881.32	n.a.	2614.66
South India	1693.22	738.18	747.04	1551.44	1850.04	2230.22	1770.95
All India	1614.16	878.64	1249.81	1385.23	1595.51	1849.32	1675.00

Source: Calculated from data provided in *Tea Statistics* (Tea Board, various years).

with the highest yield level. A worrying aspect that comes out in these tables is the significant year-to-year fluctuations in the yield levels in small gardens in many districts. Because of their small size and dependence on the large estates for processing, the yield levels could fluctuate in the small gardens, but low reliability of the data itself could be another cause of such fluctuation.

TABLE 2.27
Farm-size-wise Productivity: 2003

District/States	Up to 10.12 ha	10.12–50.0 ha	50–100 ha	100–200 ha	200–400 ha	Above 400 ha	All
Darrang	1440.80	0.00	1478.63	1711.45	1888.16	2000.48	1950.80
Goalpara	1513.37	0.00	n.a.	1250.90	1517.53	1937.70	1688.31
Kamrup	2037.74	0.00	243.90	534.45	1190.51	1671.62	1250.14
Lakhimpur	1412.70	0.00	2102.04	4125.00	2038.86	1618.92	1756.97
Dibrugarh	1397.87	988.17	1567.02	1917.04	1875.60	2136.80	1843.38
Nowgong	1320.16	0.00	1420.82	680.72	1140.17	1619.59	1428.03
Sivasagar	1199.70	1245.30	1686.92	1351.20	1293.92	1310.70	1305.41
Cachar	1029.41	150.89	215.00	422.08	947.90	1621.56	1288.58
North Cachar	n.a.	n.a.	n.a.	n.a.	1641.52	822.02	1092.03
Karbi Anglong	1741.41	760.56	556.29	701.01	n.a.	1071.43	1054.60
Assam	1340.27	958.99	1432.09	1522.35	1472.25	1760.41	1600.80
Darjeeling	n.a.	125.00	499.33	596.83	535.63	518.37	545.05
Dooars (a)	3767.33	162.68	1248.72	1645.59	1649.90	1746.44	1803.12
Terai (b)	3222.00	761.54	2346.29	3203.57	2517.94	1928.20	2602.67
West Bengal	3394.21	345.50	1072.43	1511.42	1514.82	1704.07	1770.03
Tripura	437.63	1159.60	1416.94	1120.09	1306.88	1066.80	1037.37
Bihar	551.95	1518.52	n.a.	n.a.	n.a.	n.a.	565.00
Uttarakhand	0.00	0.00	166.67	97.44	266.51	n.a.	95.85
Manipur	39.17	605.04	n.a.	n.a.	n.a.	n.a.	90.22
Sikkim	0.00	n.a.	n.a.	601.12	n.a.	n.a.	548.72
Arunachal Pr.	0.00	1388.10	2243.65	n.a.	1423.82	n.a.	1419.85
Nagaland	78.33	551.02	n.a.	n.a.	n.a.	n.a.	102.74
Odisha	n.a.	n.a.	n.a.	n.a.	490.65	n.a.	490.65
Himachal Pr.	114.46	476.19	783.22	894.23	n.a.	n.a.	257.67
Mizoram	0.00	470.59	493.51	n.a.	n.a.	n.a.	104.00
Meghalaya	0.00	578.57	n.a.	n.a.	n.a.	n.a.	321.43
Others Total	228.64	846.74	1377.59	1016.72	1112.33	1066.80	645.96
North India	1475.20	876.16	1377.14	1472.61	1474.60	1741.26	1601.14
Tamil Nadu	2027.27	836.09	1369.59	2037.54	2793.88	2620.05	2202.78
Kerala	548.23	641.87	415.14	1292.49	1423.44	2471.09	1569.29
Karnataka	2437.50	629.92	2161.14	2638.50	2658.69	n.a.	2475.56
South India	1880.49	771.72	1081.20	1725.88	2130.44	2539.63	2003.70
All India	1654.43	836.82	1307.18	1521.78	1645.90	1815.34	1690.02

Source: Calculated from data provided in *Tea Statistics* (Tea Board, various years).

Farm-size-wise Growth of Productivity

Notwithstanding the limitations of the data on farm-size-wise productivity, an attempt has been made to estimate the trends in productivity levels in different farm-sizes across time. For reasons of data availability and comparability the analysis is confined to the largest four size-classes. The findings, reported in Tables 2.28 and 2.30, suggest that at the all India level in the 1980s all the size-classes recorded positive growth in productivity, but the growth rates of yield continues to decline as we move from the lowest to the highest size-class. In south India, particularly in Tamil Nadu, however, the highest growth rate was registered in the largest size-class. In Assam and West Bengal, the growth of productivity was highest in the smallest, that is, 50–100 ha, size-class. Among the districts, productivity growth was negative among the largest size-class in Kamrup and for the second largest class in Goalpara. In districts like Kamrup, Dibrugarh, Nagaon, Sivasagar and Cachar the 50–100 ha sized garden registered impressive gains in productivity (Table 2.28).

During 1991 to 2001, productivity growth was negative for *all* size-classes under consideration here in south India and also at the all India level. In Assam it was negative in all the size-classes, except for the

TABLE 2.28
Farm-size-wise Growth of Productivity: 1981–91

District/States	50–100 ha	100–200 ha	200–400 ha	Above 400 ha	All
Darrang	2.73	2.94	1.83	0.48	1.02
Goalpara	n.a.	2.72	–2.04	1.45	0.11
Kamrup	3.14	–11.57	1.09	–0.83	–0.32
Lakhimpur	–0.65	–3.43	0.56	1.40	0.74
Dibrugarh	6.57	1.76	1.53	0.57	1.25
Nowgong	3.86	–1.27	1.05	1.60	1.64
Sivasagar	3.68	7.22	2.34	1.29	2.27
Cachar	5.36	–3.17	1.27	2.42	1.95
Assam	5.18	3.84	1.49	0.98	1.58
Darjeeling	–1.90	–1.47	1.04	–0.86	0.00
Dooars (a)	34.66	0.10	0.61	0.71	0.80
Terai (b)	4.74	–1.26	3.34	2.72	2.51
West Bengal	2.76	–1.38	0.98	0.87	1.04
Tripura	9.98	–0.50	5.56	n.a.	3.96
Others Total	14.26	–0.45	8.38	n.a.	5.89
North India	5.79	2.54	1.44	0.94	1.52
Tamil Nadu	0.76	–19.84	2.03	3.16	2.07
Kerala	0.85	2.50	2.44	4.25	3.14
Karnataka	0.19	1.07	2.92	n.a.	1.91
South India	0.90	1.49	2.37	3.70	2.56
All India	4.80	2.28	1.82	1.23	1.72

Note: Growth rates are compound growth rates.
Source: Calculated from data provided in *Tea Statistics* (Tea Board, various years).

largest size-class, in which case it was very low. At the district level, growth rate of yield in the largest size-class was negative in three districts and was insignificant in two others (Table 2.29).

In the last sub-period under study, 1998–2003, for which we have computed growth rates of yield of even the smallest size-classes, yield rate declined in the largest two size-classes in all the districts of Assam. There has been a perceptible decline in yield rate in all size-classes in Assam, with the sole exception of the small tea gardens. In fact, these small gardens have registered an impressive growth in yield in Darrang, Goalpara, Kamrup, Lakhimpur, Dibrugarh, and Sivasagar districts. As a size-class the highest decline in yield has been noticed among the 10.12 ha to 50 ha size-class followed by the class of 200–400 ha (Table 2.30). Thus while the 'crisis of low growth in productivity' seems to be all pervasive, the significant point is that it the large tea gardens in Assam that have shown the highest decline in growth of yield.

TABLE 2.29
Farm-size-wise Growth of Productivity: 1991–2001

District/States	50–100 ha	100–200 ha	200–400 ha	Above 400 ha	All
Darrang	3.82	–0.56	–0.38	–0.46	–0.53
Goalpara	n.a.	–0.16	–0.14	1.42	1.46
Kamrup	5.02	19.15	–0.09	–2.61	–0.44
Lakhimpur	6.26	–2.21	–1.09	0.84	0.27
Dibrugarh	–6.44	–0.96	–1.30	0.28	–1.58
Nowgong	0.76	–0.80	–0.06	–0.86	–0.52
Sivasagar	0.91	–4.42	–0.56	0.93	0.02
Cachar	9.09	8.09	1.64	2.07	2.53
Assam	–1.84	–1.49	–0.43	0.29	–0.39
Darjeeling	3.64	–0.12	–2.80	–0.99	–1.55
Dooars (a)	–19.41	0.42	–0.53	–0.53	–0.20
Terai (b)	–6.42	2.16	1.55	1.54	2.75
West Bengal	–0.09	1.33	–0.17	–0.20	0.65
Tripura	0.08	3.33	2.59	–5.23	0.08
Others Total	–0.81	2.83	2.80	–5.23	–1.88
North India	–1.63	–0.67	–0.43	0.13	–0.21
Tamil Nadu	–7.19	18.51	–3.17	–3.17	–0.47
Kerala	–2.22	0.94	–1.21	0.04	–0.54
Karnataka	5.58	1.14	1.43	n.a.	1.33
South India	–4.16	–1.77	–2.10	–1.29	–0.49
All India	–2.10	–0.97	–0.95	–0.03	–0.24

Note: Growth rates are compound growth rates.
Source: Calculated from data provided in *Tea Statistics* (Tea Board, various years).

TABLE 2.30
Farm-size-wise Growth of Productivity: 1998–2003

District/States	Up to 10.12 ha	10.12–50.0 ha	50–100 ha	100–200 ha	200–400 ha	Above 400 ha	All
Darrang	14.72	n.a.	2.76	–2.62	–2.47	–1.88	–2.09
Goalpara	14.33	n.a.	n.a.	–0.81	–4.63	–3.17	–1.96
Kamrup	39.21	n.a.	–2.52	–9.99	–9.65	–1.58	–5.33
Lakhimpur	23.98	n.a.	9.66	n.a.	–3.68	–6.26	–4.11
Dibrugarh	7.84	–11.69	–6.31	–3.30	–2.72	–1.45	–1.68
Nowgong	0.96	n.a.	–1.24	–4.87	–10.82	–5.29	–6.84
Sivasagar	6.68	–12.37	–0.82	–3.44	–10.88	–6.56	–6.08
Cachar	–29.85	–33.12	–28.33	–21.43	–12.47	–3.14	–6.64
North Cachar	n.a.	n.a.	n.a.	n.a.	–5.65	–11.68	–7.52
Karbi Anglong	10.36	1.00	–17.57	–4.12	n.a.	15.41	6.10
Assam	7.84	–13.57	–3.88	–3.53	–7.09	–3.17	–3.63
Darjeeling	n.a.	–9.75	2.23	3.71	–3.47	–2.64	–1.33
Dooars (a)	–18.64	–44.45	–15.65	–4.19	–5.24	–4.39	–3.60
Terai (b)	–25.79	–0.25	21.46	23.30	6.92	–4.99	5.48
West Bengal	–23.59	–22.96	–1.27	9.57	–0.38	–4.41	–0.89
Tripura	–4.31	11.48	8.34	3.59	8.99	–11.92	1.55
Bihar	50.40	–23.70	n.a.	n.a.	n.a.	n.a.	32.90
Uttarakhand	n.a.	n.a.	–11.67	–16.19	–11.82	n.a.	–26.41
Manipur	n.a.	n.a.	n.a.	n.a.	n.a.	n.a.	–10.69
Sikkim	n.a.	n.a.	n.a.	–1.84	n.a.	n.a.	–0.26
Arunachal Pr.	n.a.	43.75	32.48	n.a.	22.69	n.a.	30.20
Nagaland	n.a.	105.31	n.a.	n.a.	n.a.	n.a.	13.72
Odisha	n.a.	n.a.	n.a.	n.a.	2.81	n.a.	2.81
Himachal Pr.	–28.32	–25.92	n.a.	n.a.	n.a.	n.a.	–23.08
Mizoram	n.a.	54.01	n.a.	n.a.	n.a.	n.a.	12.16
Meghalaya	n.a.	–9.85	n.a.	n.a.	n.a.	n.a.	–22.17
Others Total	–13.60	13.61	15.29	6.66	14.76	–11.92	–1.38
North India	6.39	–8.85	–1.20	0.33	–4.57	–3.57	–2.95
Tamil Nadu	14.64	–11.32	2.79	7.93	0.90	4.56	1.47
Kerala	–1.20	–8.19	–17.02	–3.60	–6.65	0.58	–3.30
Karnataka	–11.65	–9.07	3.99	–0.50	–2.67	n.a.	–1.16
South India	14.06	–10.62	–1.16	1.96	–2.18	2.29	0.21
All India	10.12	–9.45	–1.20	0.68	–3.91	–2.88	–2.16

Note: Growth rates are compound growth rates.
Source: Calculated from data provided in *Tea Statistics* (Tea Board, various years).

Age Distribution of Tea Bushes

Another important issue relating to productivity in the tea plantation system is the age distribution of the tea plantations. The age of tea bushes has significant implications for productivity as older bushes gradually become less productive. Timely investment in expansion of area under new bushes has significant implications for sustained growth of productivity in the tea plantations. Generally, bushes over 50 years old are considered less productive. The data on age of bushes in tea plantations in Assam and other tea producing states is presented in Tables 2.31 through 2.36. In absolute terms, the total area under plantations in Assam has gone up from 193,979 ha in 1980 to 230,310 ha in 2002. During this period, the area under bushes aged more than 50 years has gone up from 65,777 ha to 67,302 ha. As a share in total plantations the area under older bushes (that is, more than 50 years of age) has changed from 34 per cent in 1980 to nearly 29 per cent in 2002, presumably because of growth of small gardens in recent years. However, this data fails to indicate the trends in the larger tea gardens separately. It is important to note that in Assam the share of

TABLE 2.31
Area under Different Age Group of Bushes: 1980

District/States	Below 5 Years	5–10 Years	11–20 Years	21–30 Years	31–40 Years	41–50 Years	Over 50 Years	Total
Darrang	8.50	10.48	17.45	11.43	6.62	9.56	35.95	100.00
Goalpara	12.56	7.90	18.99	5.32	3.76	30.65	21.07	100.00
Kamrup*	7.18	19.82	28.06	8.17	10.52	14.83	27.54	100.00
Lakhimpur*	6.19	7.37	14.31	8.38	2.98	5.01	30.46	100.00
Dibrugarh	8.87	8.45	17.05	14.86	8.77	10.14	31.86	100.00
Nowgong	4.02	6.34	16.37	19.78	15.37	13.41	24.80	100.00
Sivasagar	9.33	10.92	18.77	13.70	10.03	9.87	27.45	100.00
Cachar & North Cachar	9.09	6.62	11.34	8.32	7.20	9.78	47.65	100.00
Assam	8.78	9.29	16.55	12.67	8.58	10.22	33.91	100.00
Darjeeling	1.79	1.41	4.52	3.92	2.52	6.67	79.17	100.00
Dooars (a)	6.08	5.72	16.13	10.54	5.47	7.31	48.76	100.00
Terai (b)*	10.39	7.74	16.69	11.48	6.30	11.76	38.44	100.00
West Bengal	5.36	5.06	13.78	9.27	4.95	7.70	53.88	100.00
Others Total	4.93	5.48	8.56	12.51	6.53	15.16	46.83	100.00
North India *	7.61	9.91	15.67	11.61	7.39	9.57	40.49	100.00
Tamil Nadu	7.46	16.11	13.84	10.78	13.41	22.76	15.63	100.00
Kerala	1.22	3.20	4.96	4.37	7.67	16.92	61.66	100.00
Karnataka	4.61	3.18	4.77	2.33	21.46	45.79	17.86	100.00
South India	3.39	7.42	7.41	6.12	9.06	17.50	49.10	100.00
All India	6.75	7.78	13.83	10.48	7.73	11.19	42.25	100.00

Note: * Based on incomplete coverage.
Source: Computed from data provided in *Tea Statistics* (various years).

TABLE 2.32
Area under Different Age Group of Bushes: 1985

District/States	Below 5 Years	5–10 Years	11–20 Years	21–30 Years	31–40 Years	41–50 Years	Over 50 Years	Total
Darrang	8.15	10.00	15.37	14.18	8.85	6.62	36.84	100.00
Goalpara	12.20	11.51	16.98	11.38	5.38	10.42	32.13	100.00
Kamrup	11.01	16.48	11.12	9.61	10.94	11.36	29.49	100.00
Lakhimpur	7.46	8.48	9.29	12.28	6.65	2.59	53.25	100.00
Dibrugarh	8.42	10.61	13.49	16.48	10.51	9.40	31.10	100.00
Nowgong*	4.46	6.02	10.62	17.21	16.57	15.72	28.51	100.00
Sivasagar*	7.27	4.78	16.39	14.33	11.76	10.71	28.45	100.00
Cachar & North Cachar*	6.95	8.80	10.86	9.66	6.81	8.62	50.45	100.00
*Assam**	7.72	10.27	9.78	14.19	10.13	9.35	34.17	100.00
Darjeeling	2.88	1.67	3.06	4.03	2.98	3.63	81.75	100.00
Dooars (a)	5.65	7.07	10.66	14.23	5.78	6.98	49.63	100.00
Terai (b)*	31.84	9.60	13.32	13.99	7.40	7.30	41.58	100.00
West Bengal	5.26	6.35	9.53	12.26	5.45	6.38	54.77	100.00
Others Total*	7.55	7.17	9.31	12.93	9.94	17.86	35.22	100.00
North India	7.01	9.03	12.66	13.58	8.72	8.62	40.38	100.00
Tamil Nadu	1.63	11.17	12.77	7.48	11.24	12.59	43.12	100.00
Kerala	0.53	0.99	2.42	5.72	8.07	10.38	71.89	100.00
Karnataka	3.32	3.16	6.64	3.32	18.62	47.46	17.48	100.00
South India	1.16	6.15	7.72	6.54	9.94	12.46	56.03	100.00
All India*	0.59	8.47	11.70	12.21	8.95	9.37	43.43	100.00

Note: * Based on incomplete coverage.
Source: Computed from data provided in *Tea Statistics* (various years).

TABLE 2.33
Area under Different Age Group of Bushes: 1990

District/States	Below 5 Years	5–10 Years	11–20 Years	21–30 Years	31–40 Years	41–50 Years	Over 50 Years	Total
Darrang	7.71	9.17	15.14	15.58	10.23	5.95	36.21	100.00
Goalpara	12.83	15.05	10.57	14.51	5.75	2.19	39.10	100.00
Kamrup	9.18	12.75	11.36	8.40	6.81	9.67	41.82	100.00
Lakhimpur	9.21	10.88	6.75	13.04	6.24	3.78	50.10	100.00
Dibrugarh*	8.11	10.67	12.19	16.60	12.87	8.07	31.05	100.00
Nowgong	4.34	6.74	7.09	15.42	20.14	14.30	31.97	100.00
Sivasagar	8.38	11.48	16.76	15.79	12.61	9.83	25.16	100.00
Cachar & North Cachar	7.54	8.97	10.85	11.95	8.41	9.51	42.78	100.00
Assam	8.00	10.32	13.47	15.20	11.69	8.48	32.85	100.00
Darjeeling	3.28	2.66	2.66	4.35	2.61	2.80	81.64	100.00

(Continued...)

(Table 2.33 Continued...)

District/States	Below 5 Years	5–10 Years	11–20 Years	21–30 Years	31–40 Years	41–50 Years	Over 50 Years	Total
Dooars (a)	6.52	7.01	9.49	15.15	9.24	5.29	47.30	100.00
Terai (b)	9.25	9.11	9.73	15.12	7.68	8.09	41.03	100.00
West Bengal	6.24	6.45	8.22	13.08	7.78	5.17	53.07	100.00
Others Total	11.88	9.50	9.00	14.29	8.60	15.39	31.34	100.00
North India	7.55	9.13	11.78	14.54	10.44	7.62	38.94	100.00
Tamil Nadu	1.91	11.68	10.95	9.55	9.88	9.48	46.56	100.00
Kerala	1.58	0.82	1.82	5.99	8.29	7.06	74.45	100.00
Karnataka	2.74	3.24	5.32	6.13	11.35	35.78	35.43	100.00
South India	1.78	6.45	6.59	7.81	9.18	9.05	59.13	100.00
All India	6.42	8.60	10.76	13.22	10.19	7.90	42.90	100.00

Note: * Based on incomplete coverage.
Source: Computed from data provided in *Tea Statistics* (various years).

TABLE 2.34
Area under Different Age Group of Bushes: 1995

District/States	Below 5 Years	5–10 Years	11–20 Years	21–30 Years	31–40 Years	41–50 Years	Over 50 Years	Total
Darrang	8.05	9.25	15.44	15.14	11.46	7.16	33.50	100.00
Goalpara	13.77	14.19	15.82	16.83	6.70	4.60	28.09	100.00
Kamrup	6.47	10.04	13.40	7.29	6.57	22.63	33.59	100.00
Lakhimpur	5.66	10.72	9.26	14.65	7.17	3.96	48.58	100.00
Dibrugarh	6.40	11.03	14.07	14.14	14.25	9.08	31.03	100.00
Nowgong	7.57	5.72	8.02	11.49	17.04	16.30	33.86	100.00
Sivasagar	7.59	11.23	17.78	15.47	13.47	8.77	25.70	100.00
Cachar & North Cachar	8.69	11.36	14.74	10.57	9.62	7.28	37.75	100.00
Assam	7.48	10.59	14.98	13.98	12.57	8.60	31.80	100.00
Darjeeling	2.21	3.24	4.89	3.09	4.54	3.80	78.23	100.00
Dooars (a)	7.84	8.21	11.14	11.15	14.61	5.80	41.25	100.00
Terai (b)	10.52	11.49	11.68	11.20	10.92	10.95	33.24	100.00
West Bengal	7.28	7.83	10.20	9.86	12.52	6.13	46.18	100.00
Others Total	24.75	24.32	18.53	6.95	4.48	7.61	13.36	100.00
North India	8.00	10.36	13.90	12.71	12.29	7.95	34.80	100.00
Tamil Nadu	22.06	6.65	10.26	8.31	7.20	8.90	36.62	100.00
Kerala	11.78	1.50	1.23	2.04	3.70	9.82	69.94	100.00
Karnataka	4.34	3.81	4.10	9.62	6.77	44.02	27.35	100.00
South India	17.33	4.43	6.33	5.72	5.72	10.12	50.35	100.00
All India	10.46	8.80	11.91	10.87	10.56	8.52	38.89	100.00

Note: * Based on incomplete coverage.
Source: Computed from data provided in *Tea Statistics* (Tea Board, various issues).

Table 2.35
Area under Different Age Group of Bushes: 2000

District/States	Below 5 Years	5–10 Years	11–20 Years	21–30 Years	31–40 Years	41–50 Years	Over 50 Years	Total
Darrang	5.44	11.42	17.79	14.09	11.77	7.30	32.20	100.01
Goalpara	20.66	12.40	10.06	15.37	9.17	7.88	24.45	100.00
Kamrup	10.79	8.45	16.09	14.66	8.39	18.79	22.82	100.00
Lakhimpur	11.38	7.92	12.07	10.39	11.78	4.89	41.57	100.00
Dibrugarh	15.62	11.96	12.09	12.23	12.55	7.53	28.03	100.00
Nowgong	6.23	6.74	7.15	12.58	16.48	17.97	32.85	100.00
Sivasagar	15.84	10.44	15.36	13.83	11.93	7.96	24.64	100.00
Cachar & North Cachar	9.86	12.51	14.24	10.76	10.33	7.10	35.21	100.00
Assam	12.61	11.25	14.07	12.78	11.94	8.01	29.34	100.00
Darjeeling	5.04	9.26	4.59	5.62	7.62	7.10	60.77	100.00
Dooars (a)	8.43	9.85	10.06	9.20	13.23	7.45	41.78	100.00
Terai (b)	24.44	10.41	10.93	8.89	7.24	11.00	27.10	100.00
West Bengal	9.84	9.82	9.27	8.58	11.58	7.82	43.09	100.00
Others Total	47.78	18.16	10.53	5.06	2.76	5.11	10.59	100.00
North India	13.42	11.17	12.59	11.28	11.43	7.83	32.30	100.00
Tamil Nadu	28.55	13.33	8.58	7.06	5.75	6.91	29.81	100.00
Kerala	11.75	2.72	1.86	1.85	2.85	9.07	69.90	100.00
Karnataka	3.49	3.35	6.97	3.91	8.48	39.30	34.50	100.00
South India	21.94	9.27	6.11	5.10	4.76	8.37	44.45	100.00
All India	15.46	10.71	11.03	9.79	9.83	7.96	35.22	100.00

Note: * Based on incomplete coverage.
Source: Computed from data provided in *Tea Statistics* (Tea Board, various issues).

TABLE 2.36
Area under Different Age Group of Bushes: 2002

District/States	Below 5 Years	5–10 Years	11–20 Years	21–30 Years	31–40 Years	41–50 Years	Over 50 Years	Total
Darrang	2.53	15.64	17.81	14.18	10.67	7.60	31.57	100.00
Goalpara	5.80	20.33	9.75	13.06	11.47	11.62	27.97	100.00
Kamrup	1.46	10.38	16.27	18.79	8.01	21.12	23.98	100.00
Lakhimpur	5.59	11.79	10.91	11.55	10.39	5.41	44.37	100.00
Dibrugarh	13.62	12.94	12.61	12.02	12.98	7.69	28.14	100.00
Nowgong	0.84	8.24	8.02	12.92	17.53	17.91	34.53	100.00
Sivasagar	10.56	13.33	15.22	14.82	13.29	8.73	24.06	100.00
Cachar & North Cachar	5.35	13.57	15.30	11.54	11.21	7.79	35.24	100.00
Assam	8.81	13.48	14.39	13.17	12.41	8.51	29.22	100.00

(Continued...)

(Table 2.36 Continued...)

District/States	Below 5 Years	5–10 Years	11–20 Years	21–30 Years	31–40 Years	41–50 Years	Over 50 Years	Total
Darjeeling	3.70	10.03	4.49	4.55	8.50	7.74	61.00	100.00
Dooars (a)	4.66	13.79	10.27	9.80	12.24	9.28	39.97	100.00
Terai (b)	19.22	15.64	10.78	10.13	7.11	11.13	25.99	100.00
West Bengal	6.35	13.42	9.41	9.00	11.00	9.27	41.55	100.00
Others Total	35.59	19.74	3.40	3.62	3.21	5.57	28.86	100.00
North India	9.18	13.72	12.54	11.61	11.64	8.61	32.70	100.00
Tamil Nadu	21.12	21.43	8.51	6.34	5.84	7.08	29.69	100.00
Kerala	11.07	3.12	1.94	1.82	2.81	9.03	70.22	100.00
Karnataka	2.69	3.68	7.45	3.86	5.09	42.41	34.83	100.00
South India	17.11	14.45	6.12	4.66	4.73	8.52	44.43	100.00
All India	11.01	13.89	11.06	10.00	10.04	8.58	35.41	100.00

Note: * Based on incomplete coverage.
Source: Computed from data provided in *Tea Statistics* (Tea Board, various issues).

bushes of less than five years increased substantially from around 8 per cent during the late 1980s and early 1990s to 12 per cent in 2002. In terms of inter-state comparison it is found that the share of older bushes is higher in a number of states like Kerala, Karnataka and West Bengal than in Assam. On the other hand, among the districts of Assam, the share of older bushes was higher in districts like Darang, Cachar, Dibrugarh and Lakhimpur in 1980, while in 2002 the districts with a higher share of older bushes were Lakhimpur, Cachar and Nagaon. Thus in spite of a remarkable increase in the area under tea plantations in Assam, primarily because of the expansion of area under small tea gardens, the share of area under older bushes has not declined very sharply. Another aspect of the inter-district variations in the age distribution of bushes is that share of older bushes is relatively high in districts where the share of large tea gardens is also high. For example, our earlier analysis in this chapter notes that share of large tea gardens continues to remain high in districts like Kamrup, Nagaon, Darrang and Lakhimpur.[5] Precisely in these districts, it is found that the share of older bushes remains relatively high.

---

[5] See section titled 'Size-Class-wise Distribution of Tea Gardens' for a discussion on the changing size-class-wise distribution of area under tea gardens. The data on size-class-wise distribution of tea bushes has been presented in Tables 2.19 through 2.22.

This indicates the possibility that larger tea gardens have not been investing enough in planting new bushes. Such an inference can be conclusively proved only when we have separate data on the age distribution of tea bushes in the relatively large tea garden. In the absence of such data it is not possible to put the argument in definite terms. Our discussion with managers, particularly with those at the relatively junior level, gave us the impression that some of the tea plantations have been used by their owners as mechanisms to siphon off profit from the tea sector to other, more lucrative ventures, for a variety of reasons, including insurgency-related uncertainty, volatile market conditions and declining profits (Mishra et al. 2008). If this inference is found to be true than it implies that lack of investment in productivity enhancing technologies could be at the bottom of the crisis faced by the tea sector.

The Crisis in the Tea Industry

The tea industry in India is said to be facing a severe crisis, particularly after the disintegration of the Soviet Union, the largest importer of Indian tea. However, over the past decade, domestic consumption of tea has increased at a faster rate than production — at a steady rate of around 15 to 20 million kg annually. The steady increase in domestic demand and the inability of the tea sector to enhance production has resulted in a decline in tea exports. India's share in global export of tea has declined drastically in the recent period. The decision of the government to allow cheaper tea imports from Bangladesh and Sri Lanka, according to tea producers, has only deepened the crisis (Bhowmik 2002). However, Kumar et al. (2008) report that the fall in price of tea was observed in India as well as in other tea producing countries. Export competitiveness of Indian tea, as measured through the Export Performance Ratio, they demonstrate, fell sharply during the late 1990s, although the trend was reversed through corrective measures in the form of promotional efforts of the government in recent years.

It is interesting to note that the 'crisis' in the tea sector is not reflected in shrinkage of production or in declining prices and sale in domestic market. Domestic consumption of tea in India has gone up from an average of 518 million kg during 1990–92 to 714 million kg in 2002–4. Production of tea in the country has gone up from an average of 726 million kg to nearly 870 million kg during the same period. The crisis is better seen in loss of export share. India's share in global tea exports has declined from 17.20 per cent in 1992 to nearly 12 per cent in 2004. In absolute terms, tea export from India declined from 208 million kg in the triennium ending 1991–92, to 183 million kg in the triennium ending 2003–4. At times, the fluctuating auction prices are explained in terms of rising imports from other countries with a lower cost of production. It is true that imports have increased from nearly 1.37 million kg in 1992–93 to nearly 15 million kg during the triennium ending 2003–4, but it still constitutes only 2.4 per cent of tea consumed in India. It is doubtful whether cheaper imports could be cited as a major cause of the crisis facing the tea industry. The crisis could a result of near lack of investment, possibly because of treating the tea sector as a *residual avenue* for investment by the traditional/older firms in view of much more promising investment avenues elsewhere. While growing competition from newer foreign competitors both in global and domestic markets has led to the loss of export share, lower productivity due to lower investment could be the main contributory factor behind the crisis.

Conclusion

The analysis of the growth performance and structural features of the tea sector in Assam at a disaggregated level discussed in this chapter bring out the diverse manifestations of the current crisis in the tea industry. There has been significant expansion of small tea gardens in Assam, particularly in last two decades. But in terms of share in output their contribution remains marginal. The key findings that emerge from the foregoing analysis are listed as follows.

First, although there has been some growth in area under tea because of increase in the number of small tea gardens, in terms of growth performance in production and yield of tea, the overall picture remains grim. The growth performance of the tea sector in recent years has further worsened.

Second, Assam, as a significant contributor to tea production in India, by and large, has been at the centre stage of the worsening performance of the sector. Assam is not an exception to the crisis; in almost every aspect of its manifestation it has followed the national trend, which, of course, is hardly surprising given the share of Assam in the country's tea production.

Third, the detailed analysis of the growth of productivity in tea gardens in districts of Assam, vis-à-vis that in other parts of the country, clearly drives home the point that at the core of the poor performance of the tea industry lies the failure in raising productivity at an appreciable rate. Although the crisis per se was not the focus of the analysis, our findings on the changing growth performance on the productivity front demonstrates the pervasiveness and profundity of the problems faced by the industry.

Finally, if the crisis is one of low and declining productivity, unmistakably it is a crisis faced, more than anywhere else, in the large and medium-sized gardens of the state. No doubt the entry of large number of small tea gardens has radically altered the structure of the tea industry in the country and in Assam in particular, but as of now, the crux of the solution lies squarely in the old, large and medium-sized gardens. Because of their share in total production, and more importantly the high negative growth rates in yield in some of these gardens in recent years, they have contributed significantly to the spirals of low replanting and low yield. Added to this, changes in the age distribution of bushes at the district level seem to suggest that although the share of area under older, relatively less productive, bushes has declined at the state level between 1980 and 2002, the share of such bushes continues to remain high in those districts where there has been relatively less significant growth in the area under small gardens. This seems to suggest that older and larger plantations have not been investing enough to plant new bushes. Needless to add, all these changes have profound implications for the labour that depends on the tea sector for their livelihood.

3

Employment Characteristics
of Tea Labour

The plantation sector in Assam has evolved through different phases of regulation and government intervention. As discussed in the previous chapter, historically, the development of the plantation sector under colonialism had a profound impact on its production organisation. The two basic requirements of the plantation sector — large areas of land and a large labour force — were rarely available at the same time and at the same place. Coercion, low wages and immigrant labour were initially the three inseparable components of the plantation system and the colonial state particularly played an important role in backing the planters in their pursuit of higher profits through use of brute force. The Workmen's Breach of Contract Act, 1859, and the Inland Immigration Act, 1863, for example, virtually legalised the use of force against the indenture labourers by the planters.

After initial experiments with employing Chinese slaves and local labourers from Assam, labour in the tea gardens of Assam and Bengal were recruited from among the tribal people of central India (Guha 2006). While the initial phase of the expansion of plantation was supported by the colonial state, subsequently the sector was brought under legislation seeking to provide minimum wages and ensuring better work conditions, both by the late colonial state as well as by the government after independence. These two central features, namely, the availability of a legal framework for safeguarding the rights of workers and the predominance of the migrant labourers and their descendents as major suppliers of labour have been crucial to the functioning of the tea labour market in Assam and elsewhere. In this chapter an attempt has been made to understand the demand-supply scenario in the tea labour market in Assam, particularly in relation to the changes in employment growth and labour productivity.

Trends in Labour-use and Employment

The production process in the tea sector, particularly in the tea plantations, is highly labour intensive. The tea sector provides direct employment to around 1.26 million workers in India, apart from providing direct sustenance to another 1.23 million people, who are the dependents of the workers. Around 600,000 were employed in the tea gardens of Assam in the year 2004. In the years 2002–4, on average, 73 per cent of the total tea garden labourers in Assam were resident workers, while at the all-India level 72 per cent were resident workers. In the districts of Assam the share of resident workers varied from less than 50 per cent in Goalpara to more than 80 per cent in north Cachar and Kamrup (Table 3.1). Among the outside workers, 33 per cent were permanent workers in Assam. The districts in which the share of casual workers in total non-resident workers was high include Goalpara, Kamrup, Lakhimpur, Karbi Anglong and Cachar. There has been a steady growth of employment in absolute terms in the tea gardens of Assam. Along with increase in the volume of production, and in recent years with the expansion of area under tea gardens, there has been a greater demand for labour. The triennial average of labour employed increased from 460,000 at the end of 1982 to nearly 620,000 at the end of 2004 (Table 3.2).

To find out the extent of employment growth, and the spatio-temporal variations in it, the growth rates of average estimated labour employed in the districts of Assam as well as in major tea producing states of India, for four distinct time periods, are presented in Table 3.3. For the entire period under study, Table 3.3 shows that employment has grown at a higher rate at the all-India level than in Assam. South India as a whole and Tamil Nadu in particular has recorded substantial growth in employment during 1980–2004. In the northern states, Assam shows a higher growth rate of employment than West Bengal and all other states taken together. Within Assam, higher employment growth was witnessed in districts like Goalpara and Cachar. When the growth differentials during the two decades of the 1980s and 1990s are taken into consideration, it is clear that

TABLE 3.1

Total Labour Employed and Percentage of Resident and Outside Workers: (Triennial Averages) 2002–4

States	Districts	Total Labour Employed (as on 31 Dec)	Percentage of Resident to Total Labour	Percentage of Outside to Total Labour	Percentage of Permanent Labour to Total Outside Labour	Percentage of Temporary Labour to Total Outside Labour
Assam	Darrang	118,119	73	27	40	60
	Goalpara	9,888	53	47	6	94
	Kamrup	9,313	80	20	20	80
	Dibrugarh	202,770	75	25	40	60
	Lakhimpur	14,115	60	40	23	77
	Nowgong	19,204	64	36	35	65
	Sivasagar	160,502	73	27	33	67
	Karbi Anglong	4,791	72	28	22	78
	North Cachar	10,640	82	18	21	79
	Cachar	69,493	72	28	13	87
	Total	618,834	73	27	33	67
West Bengal	Darjeeling	52,860	89	11	65	35
	Terai (a)	39,829	60	40	19	81
	Dooars (b)	168,040	84	16	21	79
	Total	260,730	82	18	26	74
Others	Tripura	11,759	64	36	23	77
	Bihar	115	0	100	0	100
	Uttarakhand	343	78	22	75	25
	Manipur	638	0	100	0	100
	Sikkim	410	100	0	–	–
	Arunachal Pr.	1,577	80	20	0	100
	Nagaland	168	61	39	0	100

	Odisha	396	0	100	21	79
	Himachal Pr.	1,016	100	0	–	–
	Meghalaya	223	100	0	–	–
	Mizoram	87	60	40	14	86
North India		896,296	75	25	31	69
Tamil Nadu	Coimbatore	28,295	91	9	19	81
	Kanyakumari	196	72	28	0	100
	Madurai	3,303	93	7	0	100
	Nilgiris	236,162	51	49	10	90
	Tirunelveli	2,078	83	17	0	100
	Total	270,034	56	44	10	90
Kerala	Wynaad	14,031	88	12	58	42
	Idukki	64,684	91	9	80	20
	Kottayam	1,275	89	11	0	100
	Palghat	2,285	92	8	45	55
	Quilon	645	89	11	0	100
	Trichur	1,824	92	8	0	100
	Trivandrum	944	97	3	0	100
	Kerala	85,689	90	10	71	29
Karnataka	Coorg	772	87	13	0	100
	Chikmagalur	2,828	94	6	0	100
	Hassan	689	84	16	0	100
	Total	4,288	91	9	0	100
South India		360,010	64	36	14	86
All India Total		1,256,307	72	28	25	75

Source: Tea Board (2004) and for earlier years *Tea Statistics* (Tea Board 2003)

Note: Labour employed as on 31 December of each year.

TABLE 3.2

Production, Labour Employed and Production per Labour in Tea Producing Districts and States of India: 1980–2004

Districts/States	TE 1982			TE 1992			TE 2002			TE 2004		
	P	L	P/L	P	L	P/L	P	L	P/L	P	L	P/L
Darrang	606	866	700	810	1029	787	778	1162	670	797	1181	674
Goalpara	38	64	605	55	76	726	63	96	659	60	99	612
Kamrup	39	68	590	47	83	580	45	90	500	44	93	474
Dibrugarh*	1227	1667	736	1533	1914	801	1749	2145	815	1822	2169	840
Dibrugarh	1227	1667	736	1450	1773	817	1658	2004	828	1736	2028	856
Lakhimpur	0	0	0	83	140	595	91	142	640	86	141	610
Nowgong	91	143	640	120	176	681	118	190	621	115	192	597
Sivasagar#	711	1760	404	1019	2140	476	1252	2278	550	1122	2300	488
Sivasagar	707	1261	561	951	1542	617	1197	1597	750	1075	1605	669
N. Cachar	0	0	0	54	58	0	55	104	528	47	106	445
K. Anglong	0	0	0	0	17	0	19	45	429	20	48	417
Cachar	305	500	610	407	598	680	496	681	728	433	695	624
Assam	3016	4567	661	3989	5490	727	4521	6111	740	4413	6188	713
Darjeeling	130	457	285	144	471	304	97	520	185	98	529	186
Terai†	162	229	707	222	384	578	474	389	1217	595	398	1492
Dooars+	1026	1479	694	1266	1627	778	1281	1651	776	1314	1680	782
West Bengal	1316	2165	608	1532	2482	617	1851	2561	723	2007	2607	769
Tripura	32	83	388	53	112	475	65	121	535	74	118	633
Bihar	0	2	40	1	1	1024	8	1	9734	11	1	9179
Uttar Pr.	4	12	433	5	12	760	3	4	782	2	3	684
Manipur	0	0	0	0	3	77	1	5	199	1	6	175
Sikkim	0	4	64	1	3	285	1	4	275	1	4	298
Arunachal Pr.	0	0	0	1	10	116	10	17	602	17	16	1060
Nagaland	0	0	0	0	0	0	1	2	278	2	2	1057
Odisha	0	0	0	0	1	0	1	3	399	1	4	284

Himachal Pr.	4	81	47	12	10	11	10	1056	7	10	732
Meghalaya	0	0	0	0	0	1	2	325	1	2	327
Mizoram	0	0	0	0	0	0	1	358	1	1	812
Others Total	40	182	223	73	145	103	171	600	118	167	703
North India	4372	6914	632	5595	8151	6475	8843	732	6538	8963	729
Coimbatore	228	255	896	276	277	318	256	1241	300	283	1064
Kanyakumar	2	2	745	2	2	1	2	715	1	2	516
Madurai	25	26	952	30	20	30	31	952	25	33	759
Nilgiris	385	562	686	782	677	958	2356	407	1198	2362	509
Tirunelveli	17	16	1021	22	17	15	18	820	16	21	792
Tamil Nadu	657	861	764	1112	996	1322	2664	497	1540	2700	571
Wynaad	80	4	23827	100	11	137	11	12238	126	13	10054
Idukki	359	518	693	461	520	481	590	815	444	647	689
Kottayam	1	16	83	4	15	3	20	135	5	23	206
Palghat	13	24	547	17	14	20	6	3642	22	6	3382
Quilon	8	11	750	4	19	4	16	233	3	18	172
Trichur	10	15	705	15	17	17	8	2084	15	9	1580
Trivandrum	8	123	69	7	134	2	129	17	3	140	20
Kerala	481	711	677	607	729	664	779	853	618	857	724
Coorg	4	4	912	5	5	8	7	1152	8	8	1035
Chikmagalur	24	26	924	28	28	35	26	1356	36	28	1290
Hassan	7	9	732	9	8	12	6	1867	11	7	1534
Karnataka	34	39	878	42	41	54	39	1404	55	43	1284
South India	1172	1611	728	1761	1767	2040	3482	587	2213	3600	615
All India	5545	8525	651	7356	9917	8515	12324	691	8750	12563	696

Notes: i) TE = Triennium ending P=production in hundred thousand kg; L=average daily no. of labour employed in hundreds; and P/L=production per labour in kg.

ii) Dibrugarh* including north Lakhimpur district; # including north Cachar and Mikir Hills; † including west Dinajpur; + including Cooch Behar district.

iii) Major Tea producing States and Regions are italicised (Tables 3.2–3.3).

Source: Tea Statistics (Tea Board, various years).

TABLE 3.3
Growth of Employment

	Time Period				
Districts/ States	*1980–2004*	*1981–90*	*1991–2000*	*1991–2004*	*1998–2004*
Darrang	1.49	1.95	1.35	1.17	0.84
Goalpara	1.89	0.91	1.36	1.32	1.42
Kamrup	0.83	−1.37	1.47	2.03	1.23
Dibrugarh*	0.92	1.30	0.96	0.93	0.90
Nowgong	1.28	3.50	0.27	0.29	0.49
Sivasagar#	1.01	1.87	0.00	0.59	0.94
Cachar	1.51	2.96	1.85	1.38	0.75
Assam	1.43	2.06	1.08	1.06	0.87
West Bengal	0.74	1.94	0.08	0.36	0.97
Others Total	−0.58	−0.27	1.91	0.86	−0.69
North India	1.18	1.98	0.72	0.80	0.87
Tamil Nadu	5.38	1.67	10.02	10.64	10.44
Kerala	0.74	0.38	0.55	1.36	3.47
Karnataka	0.03	0.88	−1.30	0.21	3.67
South India	3.71	1.11	6.68	7.42	8.06
All India	1.76	1.82	2.02	2.29	2.53

Notes: i) All growth rates are compounded.
ii) Dibrugarh* includes north Lakhimpur district; # includes north Cachar and Mikir Hills.
Source: Computed from the data provided in *Tea Statistics* (various years).

employment growth suffered serious setbacks during the later decade in Assam, but continued to show a higher growth rate at the all-India level, owing to an exceptionally high expansion of employment in Tamil Nadu. In Assam, the only districts where employment growth improved in the 1990s, as compared to the earlier decade, were Goalpara and Kamrup; in all others, and particularly in Sivasagar and Nagaon, it declined substantially. When a relatively longer view of the employment scenario is considered, from 1991 to 2004, it is found that in Dibrugarh (which includes Lakhimpur, Nagaon and Sivasagar) growth of employment was less than 1 per cent during this period. The worrying aspect on the employment front is manifested in the fact that for the latest period under consideration, that is, 1998–2004, employment has grown at a rate of less than 1 per cent per annum in all the districts of Assam, except in Goalpara and Kamrup. Although at the all-India level, growth of labour employed has been relatively robust during this period, mainly on account of the tremendous expansion of employment in Tamil Nadu, in Assam the growth rate has decelerated to 0.87 per cent per annum. As it has already been mentioned, this slowing down of employment growth in Assam has occurred in the backdrop of increase in the area under tea as a result of expansion of area by small tea growers.

Spatio-Temporal Variations in Labour Absorption

To understand the changing dimensions of labour absorption in the tea gardens, the trends in labour-use per hectare has been analysed with respect to inter-district and inter-state variations. In 2004, labour per hectare was higher in south India than that in north India; it was remarkably high in Tamil Nadu than in other major tea producing states. Labour per hectare has been consistently lower in Assam than in West Bengal (see Table 3.4). Labour-use in Assam increased from 2.24 in 1980 to around 2.52 in 1995–97. After 1998, possibly because

TABLE 3.4
Labour per Hectare in States of India: 1980–2004

Year	Assam	West Bengal	Others Total	North India	Tamil Nadu	Kerala	South India	India
1980	2.24	2.34	1.37	2.23	2.33	1.97	2.15	2.22
1981	2.26	2.27	1.46	2.23	2.30	1.96	2.13	2.21
1982	2.19	2.25	2.16	2.21	2.35	2.06	2.20	2.21
1983	2.18	2.27	1.60	2.19	2.31	2.02	2.16	2.18
1984	2.21	2.30	1.55	2.21	2.37	2.06	2.21	2.21
1985	2.24	2.37	1.52	2.25	2.33	1.94	2.13	2.23
1986	2.16	2.31	1.52	2.18	2.30	1.89	2.12	2.17
1987	2.16	2.38	1.81	2.21	2.19	1.94	2.08	2.19
1988	2.31	2.50	1.86	2.35	2.59	2.25	2.41	2.36
1989	2.37	2.47	1.66	2.38	2.43	2.04	2.24	2.35
1990	2.35	2.46	1.63	2.36	2.64	2.15	2.40	2.37
1991	2.33	2.48	1.40	2.38	2.55	2.11	2.33	2.37
1992	2.40	2.41	8.52	2.37	2.52	2.07	2.30	2.36
1993	2.39	2.42	1.72	2.38	2.53	2.09	2.31	2.37
1994	2.46	2.59	1.49	2.47	2.34	2.06	2.21	2.41
1995	2.51	2.55	1.38	2.48	2.26	1.97	2.13	2.41
1996	2.51	2.46	1.31	2.45	2.19	2.12	2.15	2.39
1997	2.52	2.37	1.34	2.43	2.28	2.03	2.16	2.38
1998	2.34	2.34	1.19	2.30	1.80	2.01	1.88	2.21
1999	2.31	2.34	1.07	2.26	3.55	2.02	2.99	2.43
2000	2.26	2.36	1.03	2.23	3.44	2.09	2.97	2.40
2001	2.27	2.17	1.11	2.24	3.54	2.09	3.04	2.42
2002	2.30	2.20	1.13	2.27	3.65	2.15	3.13	2.46
2003	2.28	2.30	0.83	2.21	3.55	2.36	3.14	2.42
2004	2.27	2.30	0.81	2.21	3.50	2.44	3.13	2.41

Source: Computed from data provided in *Tea Statistics*, various years.

of inclusion of the small tea growers, it has gone down to 2.27 per hectare in 2004. The gap between south India and Assam has gone up substantially in the post-1998 period (see Figure 3.1). At the district level, labour employed per hectare, in 2004, was relatively high in Darrang, Lakhimpur, Goalpara and Kamrup, while districts with lower labour per hectare of tea included Dibrugarh, Sivasagar and Cachar (see Table 3.5).

So far as long-term changes in labour per hectare are concerned, between 1980 and 2004, there has been insignificant growth at the all-India level as well as in Assam, and negative growth in West Bengal. Among the districts of Assam, Dibrugarh, that includes Lakhimpur, experienced negative growth rate while in Sivasagar it was very low. In the 1980s, labour per hectare increased at a rate of 0.7 per cent in Assam, while in the next decade there was negative growth of 0.36 per cent. Both Dibrugarh and Sivasagar registered negative growth rates. During 1991–2004, apart from Kamrup and Cachar, labour per hectare either declined or grew at a very slow space. Three districts, namely, Dibrugarh, Nagaon and Sivasagar, experienced negative growth in labour per hectare, while it was as low as 0.2 per cent in Goalpara. Assam, West Bengal, other states of north India and Karnataka have registered negative growth during this period. During 1998–2004, however, there has been some recovery of growth of labour per hectare at the all-India level, primarily because of a spur of employment per hectare in south India in general and Tamil Nadu in particular. The situation in Assam and West Bengal continues to be grim. As many as four out of the seven districts under consideration registered negative growth in labour per hectare, while others experienced very low growth (see Table 3.6).

Growth in Labour Productivity

In terms of average production per labour, Assam has recorded higher labour productivity in comparison to the all-India average, except for the triennium 1990–92 (see Table 3.7). Within the districts of Assam, Dibrugarh recorded the highest labour productivity during 1982–2004. When the growth rate of labour productivity during this period is considered, Assam's labour productivity has increased at a level of only 0.4 per cent per annum. The performance of four districts, namely, Darrang, Goalpara, Kamrup, and Nagaon, has been

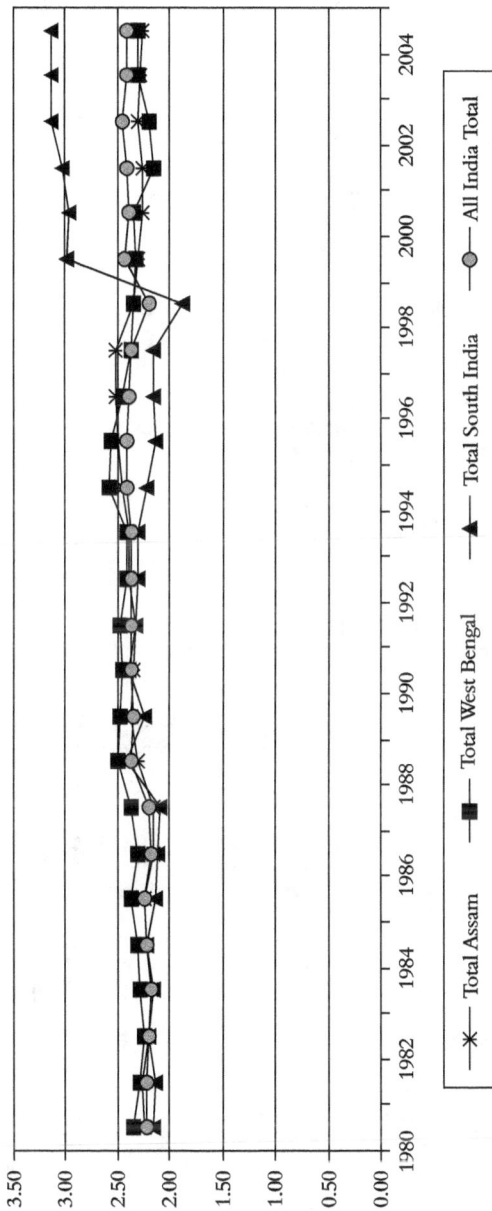

FIGURE 3.1
Trends in Labour per Hectare: 1980–2004

— ✳ — Total Assam   — ■ — Total West Bengal

— ▲ — Total South India   — ● — All India Total

Source: From data provided in Table 3.4.

TABLE 3.5

Labour per Hectare in Districts of Assam: 1980–2004

	Darrang	Goalpara	Kamrup	Dibrugarh*	Lakhimpur	Nowgong	Sivasagar	Sivasagar#	North Cachar	Karbi Anglong	Cachar	Total Assam
1980	2.46	2.12	1.88	2.59		2.20	2.15	2.15			1.52	2.24
1981	2.46	2.64	2.31	2.62		1.97	2.16	2.16			1.55	2.26
1982	2.37	2.66	2.21	2.57		1.96	2.10	2.10			1.52	2.19
1983	2.39	2.64	2.60	2.44		1.91	2.12	2.12			1.52	2.18
1984	2.38	2.63	2.98	2.47		2.14	2.13	2.13			1.59	2.21
1985	2.42	2.64	2.62	2.54		2.19	3.75	3.75			1.54	2.24
1986	2.29	1.89	2.53	2.54		1.95	2.00	2.00			1.58	2.16
1987	2.40	1.97	2.10	2.57		2.13	1.91	1.91			1.58	2.16
1988	2.71	2.21	2.04	2.63		2.49	1.95	1.95			1.90	2.31
1989	2.47	2.20	1.95	2.64		2.44	2.31	2.31			1.87	2.37
1990	2.50	2.24	1.99	2.62	3.64	2.15	2.40	2.40	0.00		1.70	2.35
1991	2.54	2.07	2.68	2.63	3.05	2.24	2.53	2.41	1.95		1.72	2.33
1992	2.66	2.87	2.19	2.66	3.21	2.35	2.52	2.38	2.06		1.67	2.40
1993	2.59	3.07	1.90	2.76	3.09	2.39	2.55	2.42	1.86	1.84	1.70	2.39
1994	2.55	3.41	2.01	2.85	3.06	2.74	2.41	2.27	1.87	2.16	1.84	2.46
1995	2.75	2.81	2.11	2.76	2.90	2.38	2.45	2.31	2.04	2.68	2.09	2.51
1996	2.71	2.46	2.14	2.82	3.68	2.42	2.28	2.18	1.44	1.71	2.43	2.51
1997	2.68	2.86	2.67	2.69	3.00	2.37	2.53	2.37	2.27	2.72	2.24	2.52
1998	2.76	2.74	2.67	2.33	2.87	2.42	3.06	2.12	14.87	2.62	2.20	2.34
1999	2.81	2.86	2.71	2.26	2.92	2.35	2.27	2.12	2.27	2.75	2.12	2.31
2000	2.79	2.75	2.59	2.16	2.92	2.37	2.23	2.10	2.58	2.41	2.12	2.26
2001	2.81	2.76	2.61	2.19	2.97	2.37	2.24	2.10	2.53	2.58	2.07	2.27
2002	2.85	2.80	2.64	2.22	3.01	2.40	2.27	2.13	2.57	2.61	2.10	2.30
2003	2.88	2.75	2.71	2.16	2.93	2.37	2.21	2.07	2.63	2.55	2.18	2.28
2004	2.85	2.74	2.73	2.17	2.84	2.35	2.20	2.06	2.67	2.55	2.17	2.27

Note: Dibrugarh* includes north Lakhimpur District; # includes north Cachar and Mikir Hills.

Source: Computed from data provided in *Tea Statistics*, various years.

TABLE 3.6
Growth Rate of Labour per Hectare

Districts/States	Growth Rate Period				
	1980–2004	1981–90	1991–2000	1991–2004	1998–2004
Darrang	0.86	0.62	1.03	0.92	0.60
Goalpara	0.91	–2.94	0.78	0.24	–0.24
Kamrup	0.68	–2.69	2.30	2.02	0.33
Dibrugarh*	–0.64	0.41	–2.34	–2.31	–1.00
Nowgong	0.79	2.14	0.08	–0.07	–0.17
Sivasagar*	0.27	0.18	–0.52	–1.10	–3.58
Cachar	1.84	2.08	3.41	1.84	0.04
Assam	0.26	0.67	–0.36	–0.59	–0.34
Darjeeling	1.32	0.90	2.47	1.62	1.09
Terai†	–0.24	4.30	–5.14	–4.67	–2.84
Dooars+	–0.13	0.67	–0.45	–0.27	0.07
West Bengal	–0.04	1.16	–0.69	–0.91	–0.56
Others Total	–2.51	0.21	–11.18	–8.57	–5.37
North India	0.12	0.83	–0.69	–0.81	–0.53
Tamil Nadu	1.66	1.11	2.00	3.89	7.58
Kerala	0.45	0.67	–0.26	0.84	3.34
Karnataka	–0.58	0.46	–1.98	–0.25	3.51
South India	1.45	0.94	1.77	3.28	6.20
All India	0.45	0.85	–0.07	0.17	1.04

Note: i) All growth rates are compound growth rates.
ii) Dibrugarh* includes north Lakhimpur district; # includes north Cachar and Mikir Hills; † includes west Dinajpur; + includes Cooch Behar district.
iii) Major Tea producing States and Regions are italicised.
Source: Computed from data provided in *Tea Statistics*, various years.

incredibly dismal on this front. However, in all these districts except Nagaon, labour productivity had increased comfortably during the 1980s. It was precisely during the 1990s that labour productivity growth slumped in many of the districts of Assam and during 1991–2004 in as many as five of the seven districts the state experienced deceleration in labour productivity growth. During the last six years, that is, 1998 to 2004, there has been a substantial deterioration in labour productivity in almost all the major tea-producing states of India. The districts of Assam have experienced a considerable decline in growth of labour productivity (see Table 3.8).

Employment Elasticity in the Tea Sector

During the 25-year period, 1980–2004, employment elasticity with respect to output in Assam is estimated to be 0.75, which is much

TABLE 3.7

Output per Labour in Districts of Assam

Districts/States	District Output per labour as % of all-India Average				District Output per labour as % of Assam's Average			
	TE 1982	TE 1992	TE 2002	TE 2004	TE 1982	TE 1992	TE 2002	TE 2004
Darrang	107	106	97	97	106	108	90	95
Goalpara	93	98	95	88	92	100	89	86
Kamrup	91	78	72	68	89	80	68	66
Dibrugarh*	113	108	118	121	111	110	110	118
Dibrugarh	113	110	120	123	111	112	112	120
Lakhimpur	–	80	93	88	–	82	87	86
Nowgong	98	92	90	86	97	94	84	84
Sivasagar#	62	64	80	70	61	65	74	68
Sivasagar	86	83	109	96	85	85	101	94
North Cachar	–	–	76	64	–	–	71	62
Karbi Anglong	–	–	62	60	–	–	58	58
Cachar	94	92	105	90	92	94	98	88
Total Assam	101	98	107	102	100	100	100	100

Notes: i) TE=Triennium Ending.
ii) Dibrugarh* includes north Lakhimpur district; # includes north Cachar and Mikir Hills.

Source: Computed from data provided in *Tea Statistics*, various years.

TABLE 3.8
Growth of Labour Productivity

Districts/ States	Growth Rates of Labour Productivity				
	1980–2004	*1981–90*	*1991–2000*	*1991–2004*	*1998–2004*
Darrang	–0.26	1.60	–1.66	–1.46	–1.32
Goalpara	0.00	2.80	–0.73	–0.95	–2.05
Kamrup	–0.33	4.22	–2.40	–2.75	–3.07
Dibrugarh*	0.37	0.25	0.33	0.71	0.62
Nowgong	0.16	–0.71	0.13	–1.06	–3.98
Sivasagar*	0.66	1.88	1.67	0.32	–0.65
Cachar	0.54	–0.23	0.66	–0.65	–4.47
Assam	0.35	0.80	0.40	–0.10	–1.62
West Bengal	0.98	–0.61	2.00	1.95	1.17
Others Total	5.77	6.65	1.78	2.86	4.92
North India	0.63	0.49	0.94	0.58	–0.67
Tamil Nadu	–1.37	5.12	–7.11	–7.10	–5.74
Kerala	0.65	2.93	0.82	–1.14	–5.12
Karnataka	2.17	1.47	4.14	2.02	–3.11
South India	–0.69	4.23	–4.46	–5.08	–5.40
All India	0.32	1.26	–0.28	–0.73	–1.79

Notes: iii) All growth rates are compound growth rates.
     iii) Dibrugarh* includes north Lakhimpur district.
Source: Computed from data provided in *Tea Statistics* (various years).

higher than that of West Bengal but lower than that of south India
(Table 3.9). As expected, employment elasticity decreased in the
1990s in comparison to the 1980s. At the district level, with the
notable exception of Goalpara, it declined for all the districts during
the 1990s. However, employment elasticity became negative in
Assam during 1998–2004. In Darrang, Goalpara, Kamrup, Nagaon,
and Sivasagar (excluding north Cachar) it has become negative.
This captures the implication of crises of low productivity in the tea
sector for labour households in Assam. It has become increasingly
difficult for them to get absorbed in the tea sector, their traditional
source of livelihood.

Changing Composition of the Tea Labour Force

Partly as a result of the labour control regimes under colonial
plantations, tea garden workers constitute male, female and child
workers. Initially, in order to ensure future supply of labour at a

TABLE 3.9
Employment Elasticity with Respect to Output

Districts/States	Employment Elasticity with Respect to Output				
	1980–2004	1981–90	1991–2000	1991–2004	1998–2004
Darrang	0.825	0.517	−0.166	−0.501	−0.101
Goalpara	0.887	0.422	2.427	1.847	−0.262
Kamrup	−0.108	−0.144	−0.166	−0.526	−0.254
Dibrugarh*	0.718	0.689	0.289	0.405	0.322
Dibrugarh	0.587	0.617	0.295	0.408	0.339
Lakhimpur	–	–	0.148	0.185	0.007
Nowgong	0.653	0.863	0.182	0.018	−0.089
Sivasagar#	0.622	0.362	0.615	0.487	0.337
Sivasagar	0.413	0.331	0.045	0.119	−0.089
North Cachar	–	–	2.783	1.349	1.461
Karbi Anglong	–	–	–	–	0.248
Cachar	0.568	0.693	0.578	0.312	−0.123
Assam	0.752	0.660	0.533	0.700	−0.377
Darjeeling	−0.252	−0.122	−0.172	−0.177	0.089
Terai	0.369	0.769	−0.007	0.049	0.145
Dooars	0.273	0.289	−0.026	0.083	−0.057
West Bengal	0.358	0.655	−0.022	0.124	0.198
Tripura	0.443	0.539	0.588	0.116	−0.182
Bihar	–	–	−0.146	0.308	0.569
Uttar Pradesh	–	–	0.429	0.323	0.228
Manipur	–	–	0.138	0.252	0.955
Sikkim	−0.080	−0.307	0.402	0.353	0.169
Arunachal Pradesh	–	–	0.334	0.213	−0.205
Nagaland	–	–	–	–	−0.265
Odisha	–	–	–	–	0.448
Himachal Pradesh	–	–	0.096	0.025	0.035
Meghalaya	–	–	–	–	0.083
Mizoram	–	–	–	–	−0.298
Others Total	−0.138	−0.134	0.574	0.240	−0.156
North India	0.627	0.691	0.295	0.457	0.057
Coimbatore	0.112	0.310	−0.046	−0.084	−1.397
Kanyakumari	−0.081	−0.613	0.463	−0.272	−0.662
Madurai	−0.025	0.094	0.302	−0.134	−0.111
Nilgiris	0.929	0.140	3.628	2.516	0.732
Tirunelveli	0.053	0.127	0.001	−0.068	0.054
Tamil Nadu	1.046	0.194	2.805	2.681	0.835
Kottayam	3.045	1.030	0.301	0.278	−0.457

Idukki	0.243	0.167	0.386	–0.243	–0.886
Palghat	0.009	–0.024	–0.021	0.020	0.241
Quilon	–1.509	–0.158	–2.783	–2.182	0.097
Trichur	–0.520	–0.366	–0.078	–0.186	–0.599
Trivandrum	–0.364	0.233	–1.618	–0.438	–0.200
Wynaad	–0.020	0.311	–0.119	0.015	–0.008
Kerala	0.258	0.182	0.152	–0.199	–1.097
Coorg	0.752	–0.116	0.379	0.503	0.793
Chikmagalur	0.054	0.888	–0.589	–0.103	1.030
Hassan	–0.646	–0.245	–0.446	–0.478	–0.960
Karnataka	0.013	0.509	–0.457	–0.122	0.485
South India	0.942	0.194	1.904	2.608	1.380
All India Total	0.779	0.520	0.817	1.234	0.049

Note: Major Tea producing States and Regions are italicised.
Source: Computed from data provided in *Tea Statistics*, various years.

reduced cost, and to discourage mobility out of the tea gardens, the planters deliberately encouraged families, rather than individuals, to migrate to the tea gardens. Usually, all the family members were encouraged to work in the gardens. Post-Independence, following the legislation that asked employers to provide housing, ration and other basic amenities to the tea garden labourers, employers now have an additional reason to encourage the participation of all family members in tea garden work. In Assam, the percentage of child and adolescent workers, however, has declined substantially from 14.12 per cent in 1980 to just around 5 per cent in 2004. This decline has been particularly sharp in the 1990s (see Table 3.10). Notwithstanding this decline the dependence on child and adolescent labour has been relatively high in Assam, in comparison to the gardens of West Bengal and south India (see Table 3.11). There are however considerable inter-district variations in the employment of child labour in Assam. For example, while in Goalpara the adolescent labourers were less than 0.5 per cent of the total labour force in 2005, in Lakhimpur district there share was as high as 12 per cent. Another noticeable feature is that between 1991 and 2004, as a result of the decline in the share of child and adolescent labour, the share of both male and female adult labourers has gone up in Assam, but the increase in the share of male labourers is far more significant than that of the female labourers.

TABLE 3.10
Changing Composition of Tea Labour Force in Assam: 1980–2004

Year	Percentage of Average Labour Employed					
	Male	*Female*	*Adolescent*	*Children*	*Adolescent+ Children*	*Total*
1980	44.21	41.67	4.33	9.78	14.12	100.00
1981	44.59	41.31	4.54	9.57	14.10	100.00
1982	44.58	41.68	4.29	9.45	13.74	100.00
1983	44.57	42.04	4.04	9.34	13.38	100.00
1984	44.63	41.86	3.94	9.57	13.51	100.00
1985	44.77	42.20	3.63	9.39	13.03	100.00
1986	44.77	42.74	3.51	8.98	12.49	100.00
1987	45.10	42.94	3.30	8.66	11.96	100.00
1988	45.63	43.15	2.96	8.26	11.23	100.00
1989	46.26	42.79	2.48	8.47	10.95	100.00
1990	46.43	43.11	2.80	7.66	10.46	100.00
1991	46.38	43.48	2.14	8.00	10.14	100.00
1992	45.70	44.56	1.97	7.77	9.73	100.00
1993	47.88	44.92	1.99	5.21	7.20	100.00
1994	48.36	45.48	2.98	3.18	6.16	100.00
1995	48.51	43.99	7.51	0.00	7.51	100.00
1996	47.68	44.14	8.19	0.00	8.19	100.00
1997	48.49	44.55	6.95	0.00	6.95	100.00
1998	48.97	44.68	6.35	0.00	6.35	100.00
1999	48.97	44.68	6.35	0.00	6.35	100.00
2000	48.74	44.93	6.34	0.00	6.34	100.00
2001	48.59	45.11	6.30	0.00	6.30	100.00
2002	48.54	45.16	6.30	0.00	6.30	100.00
2003	48.54	45.09	6.36	0.00	6.36	100.00
2004	49.57	45.19	5.23	0.00	5.23	100.00

Source: Computed from data provided in *Tea Statistics*, various years.

TABLE 3.11
Composition of Tea Labour Force: 2004

States	Districts	Percentage of Average Labour Employed			
		Male	*Female*	*Adolescent*	*Total*
Assam	Darrang	48.77	42.99	8.24	100.00
	Goalpara	54.30	45.28	0.42	100.00
	Kamrup	52.05	46.13	1.82	100.00
	Dibrugarh	50.06	46.53	3.42	100.00
	Lakhimpur	47.15	41.29	11.56	100.00
	Nowgong	49.04	44.90	6.07	100.00
	Sivasagar	47.98	45.66	6.35	100.00

	Karbi Anglong	52.07	46.99	0.94	100.00
	North Cachar	57.54	40.35	2.11	100.00
	Cachar	51.37	45.29	3.34	100.00
	Assam Total	49.57	45.19	5.23	100.00
West Bengal	Darjeeling	38.82	60.26	0.92	100.00
	Terai (a)	44.76	51.42	3.82	100.00
	Dooars (b)	49.82	47.83	2.35	100.00
	West Bengal Total	46.82	50.90	2.28	100.00
Others	Tripura	56.48	43.06	0.46	100.00
	Bihar	63.25	36.75	0.00	100.00
	Uttarakhand	50.48	48.88	0.64	100.00
	Manipur	49.78	50.22	0.00	100.00
	Sikkim	47.14	52.86	0.00	100.00
	Arunachal Pradesh	59.35	38.81	1.84	100.00
	Nagaland	56.30	43.70	0.00	100.00
	Odisha	50.77	49.23	0.00	100.00
	Himachal Pradesh	55.11	44.89	0.00	100.00
	Meghalaya	54.98	45.02	0.00	100.00
	Mizoram	52.78	47.22	0.00	100.00
	Others Total	55.89	43.60	0.51	100.00
North India		48.88	46.83	4.28	100.00
Tamil Nadu	Coimbatore	33.52	66.42	0.06	100.00
	Kanyakumari	50.00	50.00	–	100.00
	Madurai	44.81	55.19	–	100.00
	Nilgiris	47.47	52.41	0.12	100.00
	Tirunelveli	37.37	62.63	–	100.00
	Tamil Nadu Total	45.78	54.10	0.11	100.00
Kerala	Wynaad	34.43	65.53	0.04	100.00
	Idukki (c)	40.14	59.43	0.43	100.00
	Kottayam	49.01	50.99	–	100.00
	Palghat	40.22	59.78	–	100.00
	Quilon	41.56	58.44	–	100.00
	Trichur	34.77	64.57	0.66	100.00
	Trivandrum	35.84	64.16	–	100.00
	Kerala Total	39.19	60.47	0.35	100.00
Karnataka	Coorg	38.96	59.83	1.21	100.00
	Chickmagalur	44.67	54.86	0.47	100.00
	Hassan	88.78	11.22	–	100.00
	Karnataka Total	50.51	48.96	0.53	100.00
South India		44.19	55.63	0.18	100.00
All India		47.54	49.36	3.10	100.00

Source: Computed from data provided in *Tea Statistics*, various years.

Supply of Labour: Historical Antecedents and Changing Realities

Historically, migrant labourers have been the primary source for labour recruitment in the plantation sector. In its early years the Assam Company relied upon Chinese workers — some 70 workers at one stage — paying them four to five times the wage rate paid to the corresponding categories of Assamese labourers. After the services of the Chinese workers were dispensed within 1843, the local labour remained practically the sole source of labour for the industry till 1859 (Guha 2006: 12–13). Later on, labour for the tea gardens of Assam and Bengal was recruited from the tribal people of central India, namely, the Chhota Nagpur region of Jharkhand, and the neighbouring tribal belts of Odisha, Madhya Pradesh and Chhattisgarh, areas reduced to penury due to frequent droughts and famines. The policies followed by the princely states and the colonial government, which gradually eroded the communal property rights of the tribal populations over land, water and forests, also helped a great deal in forcing them to join the tea gardens in Assam and Bengal. It has been argued by Ghosh (1999) that the labour shortage, which triggered the massive recruitment drives from Chhota Nagpur and the neighbouring tribal dominated regions, was more perceived than real. It was not the non-availability of local labour but the unwillingness of the planters to employ local labour, who not only demanded higher wages, but also exercised some control over the timings and intensity of their labour-use, that led to the creation of a migrant labour force.

The planters encouraged families rather than individuals to migrate to the plantations, because, first, it ensured a dependable supply of cheap labour as labour settled in the plantations had little access to alternative occupations; second, family-based migration ensured that labour could be reproduced, thus solving to some extent the problem of future recruitment (Das Gupta 1992a, 1992b; Bhowmik 2002). The process of labour recruitment in the planta-tions of Assam did undergo significant changes but it was marked by a significant use of coercion and violence against the workers. Labour historians, most notably Behal (2006) and Behal and

Mohapatra (1992) have established the systematic use of force that was used to create the labour force and also to develop the production regime in colonial tea plantations. Behal (2006: 144) notes that the tea sector 'employed arguably the largest labour force at the lowest level of wages of any private capitalist enterprise in the history of colonial India'. Notwithstanding the fact that '[f]luctuation of tea prices in an increasingly competitive world market was a constant source of anxiety to the management of tea companies', and also that '[i]t had one of the worst records of labour relations and constantly complained of "labour troubles"', tea production and expansion of the sector continued without interruption. Arguing that the explanation of this lies in the power hierarchy that operated both at the apex level through workings of the Indian Tea Association (ITA) and also at the ground level, in the gardens, Behal argues:

> The structure of the power hierarchy based on coercion and extra-legal authority, which had dominated production relations in tea plantations for so long, began to evolve during the 1860s with the introduction of the indenture system at the height of the speculative boom during the time of the 'tea mania'... The labour force was mobilized under appalling conditions of fraudulent recruitment and insanitary transport, leading to high mortality rates and large-scale desertions from plantations. As commissioners appointed by the Government of Bengal to enquire into the affairs of the tea industry in Assam put it in their report: In the mad race of speculation, when fresh clearances were made, and acre upon acre covered with tea, to meet the terms of contract entered into with the promoters of new, or to satisfy the shareholders in old companies, no one has suffered more than the unfortunate labourer, for the opening out of new Tea Cultivation has been too often synonymous with disease and death. (ibid.: 156–57)

The colonial state not only tolerated such practices as necessary to create the tea plantations, it also took steps to legalise the use of force against the workers.

> For managers, the security of their jobs and further promotion became synonymous with ensuring uninterrupted production, and steady growth of profits for tea companies. The drive to intensify

the labour process, and constant supervision to prevent coolies from running away, reinforced those elements, described by Edgar, of physical coercion, violence, and extra-legal methods of labour control by the planters during the period of "tea mania"...Penal legislation armed the planters with immense legal powers over labour, which included the power to prosecute defaulting and malingering labourers and the power of private arrest of deserters... Extra-legal action on the part of planters was confined neither to the workplace nor to the prevention of desertions, but permeated the whole social life of a labour force isolated in the coolie lines within the plantation complex (ibid.: 160–62).

Historically, the tea planters followed a range of 'strategies to dominate, discipline, and control labour, both in work and living spaces, through legal and extra-legal methods' (ibid.: 145). In this regime of labour control, one of the key elements of disciplining labour was to restrict the mobility of tea garden labourers and to minimise their interaction with outsiders, a practice that continued up to the period of the freedom movement. Geographical isolation and surveillance, along with physical violence, were used as means to control labour. The imposition of a garden-specific work regime, under which 'the clock was put ahead by an hour from standard time to produce something known as "garden time"', was among the many measures to create a controlled industrial production organisation. It is well known that wages paid to the tea garden labourers were considered abysmally low (Savur 1973; Bhowmik 1981: 86–95; Behal, 2003). However,

> the specificity of the wage labour system in the plantations lay in the feature that even the wage earnings of all the working members of a family including child labour taken together did not allow subsistence for the family or did not make a familial wage. In the long history of the tea industry there were changes in wage rates and at times even some marginal improvements. But the above remained an *in-variant* feature (Das Gupta 1992a: PE-7, emphasis in original).

Other instruments to discipline the workers included provisioning of subsidised food as part of the wages, grant of permission to cultivate small patches of land within the garden and, in some cases, loans against obligation to work (Behal 2006). Das Gupta (1992a) has argued that the provisioning of small patches of land to workers within the gardens was a specific feature of labour relations in the plantations.

> The colonial authority as well as the capitalist planters, the latter after a period of some hesitancy, pursued an active policy of creating and maintaining subsistence economies within the tea gardens and also in the adjacent areas. They saw to it that the permanently settled immigrant labourers and their descendents had some access to means of subsistence outside the capitalist sector-paddy land, vegetable plots, grazing areas, forest produce, etc. (ibid.: PE-7-8)

One implication of the continued existence of this subsistence economy was that

> ...only a part of the cost of reproduction of the labour force was met by capital. A considerable proportion of the subsistence goods and services required for reproduction was produced by labour that was external to the capitalist production process. In other words, the cost of reproduction of labour power was, to some extent, externalised to a different economy-a non-capitalist economy and, in fact, was borne by the labourers themselves (ibid.: PE-7-8).

Since 'the surplus generated by the plantation labourers was actually extracted in three different forms — absolute surplus value, relative surplus value and labour rent and, further, the different methods of surplus-extraction worked in an interlocked fashion', ... '[t]he labour force remained semi-prolitarianised — essentially half-peasant, half- industrial wage worker' (ibid.: PE-8-9).

After independence, the Industrial Disputes Act, 1947, the Factories Act of 1948 and the Plantations Labour Act, 1951, to some extent, increased the bargaining power of the tea garden labourers and improved their working and living conditions. The Plantations Labour Act contains several provisions related to housing conditions, health and hygiene, education and social welfare (see Appendix Ia) It is, however, widely known that even today the basic requirements of the act have not been fulfilled. Lahiri (2000), for example, reports:

> Since 1947, the wage of the tea plantation labour has increased only numerically, there has been no rise in their real wage...The literacy rate among tea garden workers and their families is a poor 20 per cent. Around one third of the workforce is denied housing facilities. Every

year, hundreds of people in the plantations die from water-borne diseases like gastro-enteritis and cholera. Most of the plantations have no portable drinking water facilities and drainage systems.

Some of the key organisational features of the tea industry, which have a bearing on the labour process in the plantations, are as follows. First, the plantation system has a distinct, vertical work hierarchy that maintains the class structure of workers and management (Bhowmik 1981, 2002). A study by Krishnakutty et al. (1993) reveals that tea companies have the highest personnel cost as per cent value of production as well as total income ratio. Between 1987–88 and 1991–92, the study based on data from 248 tea companies shows that while wage costs have fallen, despite wage increases, the cost of management has increased (quoted in Bhowmik 2002). Critics point out that, 'in essence, the industry has still maintained the feudal relations of production in its pre-marketing phases' (Lahiri 2000; Raman 2010: 164–65). Second, women constitute over 45 per cent of the total labour force in the industry.[1] This is the only industry in the organised sector that employs such a high proportion of female labour. The female workers are mainly employed in plucking of tea leaves and in light maintenance work. Third, employment of adolescent and child labour as permanent workers is a special feature of the plantation industry. Non-adult permanent workers accounted for around 3.10 per cent of the total permanent workers in the tea industry as a whole in 2004. The actual incidence of child labour is considered to be higher, as many of them are employed as casual workers. Also, the incidence of child labour was found to be much higher in Assam (5.23 per cent) than in the south India (Table 3.11).[2] Adolescents are engaged in more or less the same work as the adults but they are also engaged in spraying of pesticides, which

---

[1] This has its roots in the migration labour from eastern and central India to the gardens. Movements of families were specifically encouraged during the early phases as it reduced search and transportation costs in future. Because of such shifting of families to the tea gardens, labour could be reproduced without adding substantial costs to the employers. Also see Tables 3.10 and 3.11.

[2] Although the percentage of child and adolescent workers has declined substantially from 14.12 per cent in 1980 to just around 5 per cent in 2004, the dependence on child and adolescent labour continues to be high in Assam.

may involve health hazards. Although children are supposed to be given lighter work, often in practice they are engaged in similar work as the adults. It is important to note that the bulk of child labour is employed in Assam (81.98 per cent of total permanently employed child labour in the industry in 1997). It is almost negligible in the two southern tea-growing states. Fourth, in comparison to other sectors, the rate of unionisation is fairly high in the tea industry. In Assam, where the largest number of tea plantation labourers is employed, the INTUC-affiliated Assam Cha Mazdur Sangha is the one major union, which represents workers. The effectiveness of these unions in securing and safeguarding the interests of the workers has not been very impressive. Fifth, wages of tea plantation labourers are the lowest in the organised sector. Within the tea sector, workers in Assam and West Bengal receive lower wages than their counterparts in Tamil Nadu and Kerala.

Partly because of the nature of work organisation in the tea plantations, the occupational mobility among tea garden labourers has been fairly low. The legacy of the indenture system ensures that the entire family of workers — male, female and children — tend to get employed in the same garden. The concentration and isolation of the tea garden labourers also puts effective barriers in terms of access to alternative sources of earnings. It is always in the interest of the management to ensure that the labour households remain completely dependent upon their earnings from the tea garden, as recruitment of new labour is costly. According to the existing legal provisions labourers have the right to housing. Recruitment of workers from existing labour families put no extra burden on the management, but employment of new labourers may raise that cost substantially. The employment of children in tea gardens, inadequate schooling facilities and lack of scope for alternative skill-formation lead to perpetual dependence of the children of tea garden labourers in the tea industry itself. Again, the payment of part of the wage in terms of subsidised foodgrains might act as a strong incentive not to leave the garden. The ethnic identity of the immigrant tea garden labourers also acts as an additional barrier against their occupational mobility.

Collective Bargaining and Wage Determination

In the 1840s and early 1850s the wage rate in the plantations of Assam generally varied between ₹2.50 to ₹3.50 per month. It was only after the revolt of 1857, that it rose to ₹4. In the given context, these wages were not unjust, as indentured labour did not exist till then. This was mainly because of the strong bargaining power of the local labour force (Guha 2006).

The situation changed thereafter when indentured labour became the dominant mode of labour recruitment. The labour policy that was pursued by the planters and their governments did not encourage free movement of labour at competitive wages. The planters adopted coercion as the prime mechanism for ensuring labour supply. Their methods included extensive use of force to restrict free movement of labourers and it resulted in the creation of a type of serfdom, which was unheard of in any other public works departments at that time. A common free labourer earned ₹7 per month in 1864, while in the plantation sector they woulds earn between ₹4 to ₹5 during the same period. The Transport of Native Labourers Act of 1863 did not stipulate the minimum wage (ibid.). Statutory wages were laid down by bringing amendments to existing legislation.[3] Though the provision was formally abolished in 1870 by another amendment, it was re-enforced under the Inland Emigration Act of 1882, and the same statutory wages continued in practice up to 1901. The minimum wages so set were ₹5 and ₹4 respectively for men and women and ₹3 for children per month. Under the mounting pressure of public opinion in India in support of increase

---

[3] Das Gupta (1992c: 181) observes: 'The Workmen's Breach of Contract Act (or Act III) of 1859, reinforced by Section 492 of the Indian Penal Code, introduced provisions for arrest and punishment by the government under civil laws of workers considered responsible for any breach of contract. The Act VI of 1865 (an amended version of the transport of Native Labour Act of 1863) along with certain so-called welfare provisions relating to work period, working hours, wages and sanitation, not only contained penal clauses and made deserting or 'absconding' workers criminally punishable, but also gave planters the power to arrest such workers without any warrant'.

in labourers wages, a bill was brought before the Imperial Legislative Council in 1899 to alter the time-old statutory wages for men from ₹5 to ₹6 and for women from ₹4 to ₹5. However, even this modest proposal was not acceptable to the British capitalist. The proposed rates were modified to a graduated scale, with annual increments reaching the full rates only in the fourth and last year of the contract. The average monthly wage earning of ₹8.09 for men and ₹7.59 for women, including diet rations, subsistence allowance and bonus in 1917–18 showed a slight improvement as compared to 1905–6 (DPI of Assam report, 1917–18, cited in Guha 2006). But in real terms much of it was eroded by the rising prices (ibid). There were strides against the planters' labour squeeze policy till independence.

Post-Independence, all wages and provisions of the plantation workers are being assured by bilateral agreements between the ACMS, the workers representative, and by the employers' associations.[4] The various bilateral agreements of 1966, 1969, 1970, 1973, 1974, 1976, 1980, 1983, 1993, 1996, and 2000 have from time to time revised the wages of the plantation sector workers. It was decided that fair wage on a timely basis would be guaranteed to all pluckers irrespective of the quantity of leaf plucked, provided they work a full day. However, this commitment was tempered with by fixing the quota to be fulfilled. Though the wage was specified, workers had to pluck the minimum assigned task of 27 kg of tea leaves a day. The ACMS has always been prompt to take up the matter of wage revision and the wages of the workers are revised every three years. In the tea plantation sector there is timely revision of wages through bilateral agreements. The agreements have not only fixed the minimum task of a job in the plantation but have also raised the wages linking it to the consumer price indices. All agreements and contracts are invariably wage contracts rather than output contracts. However, the diversity in contract is not explicit in the plantations of Assam. In 2001–2, for example, a worker got about ₹45 per day; therefore if the worker

---

[4] This section draws heavily on the summary of negotiations provided in Das (2002).

worked for all 26 working days in a month then he/she earned ₹ 1,170 a month. Apart from the economic benefits the workers in the tea estates also receive many non-economic benefits. In the plantation sector, tea garden workers are paid bi-monthly, after 12 days of work. The workers are *not* paid for Sunday as it is a holiday for them.

The varied wage structure in the plantations and the diversity within the regions is partly captured by Tables 3.12 and 3.13 drawn from a report of a committee set up by the Government of India (GoI 2007). The committee, which was set up by the Ministry of Commerce and Industry to recommend changes in the Plantations Labour Act, 1951 (see Appendix Ia), has reported the total labour cost to employers on the basis of estimates provided by the Tea Board.

TABLE 3.12
Wage Structure in Tea Plantations: North India

Plantation districts/States	Daily Wages (₹)		Agreement Date
	Adult	*Increment*	
North India			
Assam Valley	51.1	w.e.f. 01.11.06 – 2.60	30.11.2005
(From 01.11.2005 to 31.12.2009)		w.e.f. 01.15.07 – 3.70	
		w.e.f. 01.09.08 – 3.70	
Cachar	46.25		
(From 01.04.2005 to 31.05.2006)			
Darjeeling	48.4	1st Year – 2.50	25.07.2005
(From 01.04.2005 to 31.03.2008)		2nd Year – 2.50	
		3rd Year – 3.00	
Dooars	48.4	1st Year – 2.50	25.07.2005
(From 01.04.2005 to 31.03.2008)		2nd Year – 2.50	
		3rd Year – 3.00	
Terai	48.4	1st Year – 2.50	25.07.2005
(From 01.04.2005 to 31.03.2008)		2nd Year – 2.50	
		3rd Year – 3.00	
North Dinajpur	48.4	1st Year – 2.50	25.07.2005
(From 01.04.2005 to 31.03.2008)		2nd Year – 2.50	
		3rd Year – 3.00	
Tripura		—	
2004	26.05		
2005	28.5		
2006	30		
2007	30		

Source: GoI (2007).

TABLE 3.13
Wage Structure in Tea Plantations: South India

States	Worker	Basic (₹)	D.A. (₹)	Total (₹)	w.e.f.
Kerala	Adults	34.72	43.32	78.04	
	Adolescent	26.04	43.32	69.36	01.04.2005
Karnataka	Adults	(The DA up to 2703 points		71.00	
	Adolescent	have been merged with		63.90	01.04.2005
Tamil Nadu		basic wages)			
Nilgiris		₹ 72/– (Plus Attendance Bonus of ₹2/–)			
Nilgiri–Waynad		₹ 73/– (Plus Attendance Bonus of ₹1/–)			01.10.2004
Anamallais		Consolidated pay of ₹.72/ per day and there is			01.01 .2002
		no time limit prescribed. The wage matter is pending			
		before the Special Industrial Tribunal, Chennai.			

Note: In addition to the aforesaid daily wages various other incentives are also available to the tea plantation workers including female workers by way of plucking incentive, guaranteed time rate incentive etc., as per the general agreement, the details of which are not readily available and/or compiled.

Source: GoI (2007)

Wages in North India are significantly lower than that in the South. This difference is explained by the fact that unlike in the plantations in south India, workers in North India get their wages partly in kind.

The foodgrains provided at subsidised rates to the labourers in the gardens are supplied by the Food Corporation of India (FCI) to the gardens at a pre-determined rate, that is equal to the rate applied to families considered Above the Poverty Line (APL). The cost of transport and distribution is borne by the employers. Since, at times the foodgrains supplied through the FCI are not enough, employers have to resort to open-market purchases and hence it is considered that 'the effective procurement price varies due to the variation in the prevailing market price and in general, such price is higher than the FCI price' (ibid.: 19). As per the report, at present tea garden owners are supplying foodgrains at 54 paise per kg in Assam, 40 paise per kg in Dooars and Terai and 63 paise for rice and 54 paise for wheat in Tripura. Assuming a typical labour household as consisting of three consumption units (two adults and two children), the report, following the suggestions by the ITA, estimates that 'each garden worker is eligible to receive 10.58 kg {3. 26+ (2 × 2.44) + (2 × 1.22)} of foodgrains each week at a concessional price. The amount for receiving 10.58 kg of foodgrains paid by the tea garden workers in each week to the tea garden employers is presented in Table 3.14.

In Assam and West Bengal, a part of the 'in kind' wages also includes free firewood. While in Assam, firewood entitlement is 228 cft per annum per family, in West Bengal, the fire wood entitlement is 156.25 cft per annum per family or 715 kg of coal per annum per family. The committee estimated the cost of these entitlements in the following manner:

> The price of 228 cft fire wood in the State of Assam is reportedly ₹1350/-. In West Bengal the price of coal per kg is reportedly ₹2.55 per kg. Considering that each family may have two tea garden workers and considering the above reported price, the price of fuels supplied freely by the tea garden owners to the tea garden workers works out to ₹2.17 per worker per day in the State of Assam and ₹2.92 per worker per day in the State of West Bengal'. (GoI 2007: 20)

Following the committee's methodology, the wage rates in north India paid to tea garden workers with consideration of supply of foodgrains at a concessional price and/or fuels free of price for the year 2006 is estimated to be as low as ₹48 in Tripura. In Assam it was between ₹60 to ₹68 per day (Table 3.15). Based on these estimates, the committee arrived at estimates of labour costs borne by employers per worker per day in various regions in north India. According to these estimates, labour cost, taking into account wages, subsidised fuel and foodgrains, turns out to be ₹71 in Assam (see Table 3.16). It is important to note that even after all these adjustments, the wages in Assam, West Bengal and Tripura are lower than that in south India.

TABLE 3.14
Value of Foodgrains Paid to Workers

States	Price paid to the workers for foodgrains per week per worker (in ₹)	Price paid to the workers for food grains per worker per day (based on six working day in a week) (in ₹)
Assam	5.71	0.95
West Bengal	4.23	0.71
Tripura	6.19	1.03

Source: GoI (2007).

TABLE 3.15
Estimated Daily Wage Rates of Tea Garden Workers: 2006

Variables	Assam Valley	Cachar	Dooars & Terai	Tripura
Settlement daily wage rate (in ₹) per worker	54.80	46.25	53.90	30.00
Add Daily estimated cost of supply of foodgrains per worker (in ₹)	14.63	14.63	14.11	18.59
Daily wage including cost of foodgrain without concession component per worker (in ₹)	69.43	60.88	68.01	48.59
Less estimated amount received daily by employer from worker towards supply of foodgrain per worker (in ₹)	0.95	0.95	0.71	1.03
Estimated Daily Wage Rates*	68.48	59.93	67.3	47.56

Note: * Authors' calculation.
Source: Based on GoI (2007).

TABLE 3.16
Estimated Labour Cost in Tea Gardens Per Worker Per Day: 2006

Variables	Assam Valley	Cachar	Dooars & Terai	Tripura
Daily cost of foodgrain borne by the employer per worker (in ₹)	13.68	13.68	13.40	17.56
Daily wage including cost of foodgrain borne by employer per worker (in ₹)	68.48	59.93	67.30	47.56
Daily cost of fuel supplied freely per worker by employer (in ₹)	2.17	2.17	2.92	Nil
Daily cost of foodgrain and fuel borne by employer per worker (in ₹)	15.85	15.85	16.32	17.56
Total daily wage including cost of supply of foodgrain and fuel borne by employer per worker (in ₹)	70.65	62.10	70.22	47.56
Estimated production of made tea per labour based on 300 working days in a year (in kg)	2.41	1.85	2.66	2.07
Estimated average daily number of labour per hectare	2.28	2.17	2.17	1.39
Per kg cost of foodgrain and fuel borne by employer (in ₹ per kg of made tea)	6.58	8.57	6.14	8.42
Daily cost of foodgrain and fuel borne by employer (in ₹ per hectare)	36.14	34.39	35.41	24.41

Source: GoI (2007).

The total annual amount borne by the employer in Assam, West Bengal and Tripura (based on labour, production, area for the year 2004) is presented in Table 3.17.

It is important to note that the committee recommended that 'part wages as rations as is the practice in North India should continue', while wage structure should continue to be negotiated through bipartite or tripartite negotiations, 'but the minimum wages should not be less than the minimum wage fixed from time to time' (GoI 2007: 37). The methodology adopted by the committee, however, does not take into account the differential entitlements of permanent and casual workers. So far as the actual amount of subsidised kind wages/entitlements is concerned, labour households faced a number of additional constraints. As our field survey results reported in subsequent chapters suggest, the kind-component of the wages is influenced by a number of additional variables such as: the number of permanent and casual workers in the family, the frequency and regularity with which the non-permanent workers manage to get employment, and, above all, the seasonality associated with the tea leaf plucking activities.

TABLE 3.17
Estimated Annual Cost of Supplying Foodgrains in Tea Gardens

	Assam	*West Bengal*	*Tripura*
Estimated average daily number of labour	617,518	262,672	11,518
Production of made tea (000 kg)	435,649	214,541	7,168
Area in Hectare	271,768	114,003	8,268
Per kg cost of foodgrains borne by employer (₹ per kg of made tea)	5.68	5.04	8.48
Daily cost of foodgrain (in ₹ per hectare)	31.19	29.08	24.41
Daily cost of foodgrain borne by employer (in ₹ per worker)	13.68	13.40	17.56
Estimated total amount to be borne by the employer based on per kg basis (₹ in crore)	247.45	108.13	6.08
Estimated total amount to be borne by the employer based on per worker basis (₹ in crore)	263.57	109.82	6.31

Source: GoI (2007).

Conclusion

The demand-supply scenario in the tea labour market in Assam, particularly in relation to the changes in employment growth and labour productivity, discussed in the preceding sections brings out the problems faced by the tea sector on the employment front. For the period 1980–2004, employment has grown at a higher rate at the all-India level than in Assam. Although at the all-India level, growth of labour employed has been relatively robust during 1998–2000, mainly on account of the tremendous expansion of employment in Tamil Nadu, in Assam the growth rate has decelerated to 0.87 per cent per annum. During 1998 and 2004, however, there has been some recovery of growth of labour per hectare at the all-India level, but the situation in Assam and West Bengal continues to be gloomy. As many as four out of the seven districts in Assam under consideration, registered a negative growth in labour per hectare, while others experienced very low growth. Although labour productivity had increased relatively comfortably during the 1980s in Assam, during the 1990s labour productivity growth slumped in many districts in Assam. During the last six years, from 1998 to 2004, there has been substantial deterioration in labour productivity in almost all the major tea-producing states of India.

Employment elasticity decreased in Assam in the 1990s compared to the 1980s. At the district level, it has declined for almost all the districts during the 1990s. Further, employment elasticity was negative in Assam from 1998 till 2004. In Assam, the percentage of child and adolescent workers has declined substantially from 14.12 per cent in 1980 to just around 5 per cent in 2004, the dependence on child and adolescent labour continues to be high in Assam when compared to the gardens of West Bengal and south India. Another perceptible feature is that between 1991 to 2004, as a result of the decline in the share of child and adolescent labour, the share of both male and female adult labourers has gone up in Assam, but the increase in the share of male labourers is far more noteworthy than that of the female labourers. All these features of the labour market capture the implication of crises of low productivity in the tea sector for labour households in Assam. Increasingly they find it difficult to get absorbed in the tea sector, their traditional source of livelihood.

The uniqueness of the labour market scenario in the tea gardens lies in the fact that the supply side is governed by a set of factors that import the characteristics of a reserve army of labour. In contrast, the demand side of the labour market is highly responsive to the market conditions. The market fluctuations in labour demand conditions, our analysis indicates, have largely been absorbed by one segment of the labour force.

4

Occupational Mobility among Tea Garden Labourers

The study of intergenerational occupational mobility has received a great deal of attention from social scientists presumably because of the link between such mobility to wider notions of 'equality of opportunity' and 'openness' of the society[1] (Miller and Volker 1985).

Generally, the study of intergenerational occupational mobility is done through the analysis of simple two-dimensional matrices with categories of father's occupation arrayed across one dimension, and categories of son's occupation arrayed across the other (Matras 1960).

Suppose the initial occupational distribution (first generation) is given by a vector,

$$\alpha = (a_1, a_2, \ldots a_n) \qquad \ldots (1)$$

in which a_1 is the proportion of the population of the i-th occupation group, and

$$0 \le a_1 \ge 1; \sum_i^n a_1 = 1$$

Then the mobility pattern can be denoted by the $n \times n$ matrix:

$$A = [p_{ij}]$$

---

[1] Among the early studies, Morris Ginsberg (1929) in a paper in the *Economic Journal* discovered the extent to which sons moved out of their father's social class. In 1913, Chapman and Abott, in a paper on 'The Tendency of Children to enter their Father's Trades' in the *Journal of the Royal Statistical Society*, made a similar inquiry (Saunders 1931). Also among the early contributors to this body of literature was Saunders (1931) and De Jocas and Rocher (1957).

with
$$0 \le p_{ij} \ge 1; \sum_i^n p_{ij} = 1$$

In which p_{ij} is the proportion of those whose fathers were in the i-th occupational group and who themselves are in the j-th occupational group.

For example, suppose fathers and sons can have either of the two jobs. A matrix that summarises intergenerational mobility in location P has the following form.

$$P = \begin{bmatrix} p_{11} & p_{21} \\ p_{12} & p_{22} \end{bmatrix}$$

with father's occupation as columns and son's occupation as rows. The upper left term (p_{11}) is the number of sons of job 1 fathers who themselves obtained job 1. One straightforward and simple measure of the overall mobility in P is the proportion of sons who have taken up jobs different than their fathers.

i.e.
$$M_p = (p_{12} + p_{21})/(p_{11} + p_{21} + p_{12} + p_{22})$$

or to generalise,

$$M_p = {\sum_{i \ne j} P_{ij}} \Big/ {\sum_i P_{ij}}$$

Although M_p has the advantage of simplicity as a benchmark, its main shortcoming is that it cannot be used to compare mobility across two matrices (Long and Ferrie 2005). The drawback of this measure is that it does not distinguish between differences in mobility (*i*) arising out of differences across the matrices in the distributions of father's and son's occupations, termed '*prevalence*' (Hauser 1980); and (*ii*) arising on account of differences across matrices in the association between father's and son's occupations that may occur even if the distributions of father's and son's occupations were identical-differences in what Hauser terms '*interaction*' (Long and Ferrie 2005).

Extent and Pattern of Intergenerational Occupational Mobility

Before moving into the discussions on intergenerational mobility patterns, the following sections discuss the occupational distribution of adults in the surveyed households.

The Occupational Distribution of Workers

The distribution of workers across occupations is considered to be the result of household-level and individual responses to opportunities as well as constraints. For the purpose of intergenerational analysis of occupational mobility the occupational status of all members of the households aged 15 or above is considered.[2] We present separate analysis for labour households residing within the tea gardens and those staying outside the gardens. It was noticed that for a variety of reasons, tea labour households depend on a number of livelihood sources outside the tea gardens, although employment in the tea garden remains their mainstay. In total, 33 per cent of persons above the age of 15 work as permanent tea garden labourers. Nearly 25 per cent work as casual labourers in the tea gardens. A miniscule 3.3 per cent work as lower level supervisors and merely 0.2 per cent work at the clerical level. Thus tea gardens as a whole employ nearly 61 per cent of those adult members living within the tea gardens (Table 4.1). Among the districts the pattern is remarkably similar although in Lakhimpur the dependence on tea gardens is higher than the other two districts. Nearly 11 per cent of the adult members were unemployed at the time of the survey. So far as employment outside the tea gardens is concerned, it was found that nearly 14 per cent of the total adults were employed in non-farm casual work. The significance of such non-farm casual work of unskilled variety was much more pronounced in the Sivasagar and Dibrugarh districts than in Lakhimpur where, incidentally, agricultural labour has a higher

---

[2] In the tea gardens it was not unusual to find children entering the workforce at the age of 12. But in order to get a clearer picture of the intergenerational occupational mobility, we have taken only the persons aged 15 or above as a worker in this analysis.

TABLE 4.1
Occupational Distribution of Current Workers (Within Tea Gardens)

Sl. No	Occupations	District Code 1	2	3	Total
1	Tea Garden Casual Labour	84	96	127	307
		(24.1)	(21.6)	(28.7)	(24.8)
2	Tea Garden Permanent Labour	110	144	152	406
		(31.5)	(32.4)	(34.4)	(32.8)
3	Tea Garden Lower Supervisory Staff	15	15	11	41
	& Service Providers	(4.3)	(3.4)	(2.5)	(3.3)
4	Tea Garden Clerical Staff	0	1	1	2
		(0.0)	(0.2)	(0.2)	(0.2)
5	Cultivators	21	18	15	54
		(6.0)	(4.0)	(3.4)	(4.4)
6	Agricultural Labourers	0	10	21	31
		(0.0)	(2.2)	(4.8)	(2.5)
7	Non-farm Casual Labour (unskilled)	66	65	40	171
		(18.9)	(14.6)	(9.0)	(13.8)
8	Non-farm Casual Labour (skilled)	1	16	2	19
		(0.3)	(3.6)	(0.5)	(1.5)
9	Trade and Business	4	3	2	9
		(1.1)	(0.7)	(0.5)	(0.7)
10	Government Service	3	3	4	10
		(0.9)	(0.7)	(0.9)	(0.8)
11	Private Salaried Service	0	0	1	1
		(0.0)	(0.0)	(0.2)	(0.1)
12	Unemployed	40	62	39	141
		(11.5)	(13.9)	(8.8)	(11.4)
13	Others	0	1	0	1
		(0.0)	(0.2)	(0.0)	(0.1)
14	Non-response	5	11	27	43
		(1.4)	(2.5)	(6.1)	(3.5)
	Total	349	445	442	1236
		(100.0)	(100.0)	(100.0)	(100.0)

Note: Figures in parentheses refer to percentages to column totals.
Source: Field Survey (Tables 4.1–4.22).

presence. Skilled non-farm employment was not of much importance in overall terms, but it is significant to note that Dibrugarh had a relatively higher incidence of such employment than the other two districts. It is important to note that occupations such as trade and business, government service and private salaried service had negligible presence in the occupational distribution of adults in the tea gardens.

This simple analysis reinforces the limited occupational diversification among the tea garden labourers, particularly the limited access to better occupations.

The gender dimension of the occupational distributions has been captured through the male-female differences in the share in different types of occupations (Table 4.2). It is clear that within the tea garden occupations, women have a greater share among the casual labourers than among the permanent workers. Their presence in the relatively better-paid categories of tea garden workers such as service providers and lower supervisory staff is negligible. The striking aspect of the *gendered* process of occupational diversification, as captured through our survey, is that while women of tea garden households have a relatively lower presence in occupational categories such as cultivators and agricultural labourers, they have a substantial presence in unskilled, non-farm casual employment. Needless to add, like their male counterparts, women in tea gardens have remarkably low presence in trade and business, government or private salarised employment. Such a gendered pattern of occupational diversification could have been the result of livelihood diversification strategies within the households, which are deeply embedded in the intra-family distribution of work, power and skills.

So far as workers residing outside the tea gardens is concerned, the dependence on the tea gardens is relatively less, but not entirely insignificant. In total, 15.2 per cent of the adults in selected households outside the tea garden work in the tea gardens as casual workers, while around 28 per cent depend on the gardens for their livelihoods (Table 4.3). A majority of those who live outside the gardens are dependent on cultivation. Of all the adult workers, 26.2 per cent work as cultivators and 12.4 per cent as agricultural wage labourers. The limited access of these families to non-agricultural occupations gets manifested very clearly through the pattern of occupational distribution. Only 3 per cent of adults were engaged in trade and business, while a miniscule 1.3 per cent were in government service. The percentage of adults reported to be unemployed was, however, much higher among the 'outside the tea garden households' than those residing inside the gardens. Important inter-district differences in the occupational distribution of adults in households are as follows: First, a relatively greater dependence on or access to tea garden jobs

TABLE 4.2
Occupational Distribution of Male and Female Workers (Within Tea Gardens)

Sl. No	Occupations	Workers		
		Male	*Female*	*Total*
1	Tea Garden Casual Labour	114	193	307
		(18.6)	(31.0)	(24.8)
2	Tea Garden Permanent Labour	232	174	406
		(37.8)	(27.9)	(32.8)
3	Tea Garden Lower Supervisory Staff &	36	5	41
	Service Providers	(5.9)	(0.8)	(3.3)
4	Tea Garden Clerical Staff	2	0	2
		(0.3)	(0.0)	(0.2)
5	Cultivators	44	10	54
		(7.2)	(1.6)	(4.4)
6	Agricultural Labourers	25	6	31
		(4.1)	(1.0)	(2.5)
7	Non-farm Casual Labour (unskilled)	24	147	171
		(3.9)	(23.6)	(13.8)
8	Non-farm Casual Labour (skilled)	18	1	19
		(2.9)	(0.2)	(1.5)
9	Trade and Business	5	4	9
		(0.8)	(0.6)	(0.7)
10	Government Service	5	5	10
		(0.8)	(0.8)	(0.8)
11	Private Salaried Service	0	1	1
		(0.0)	(0.2)	(0.1)
12	Unemployed	87	54	141
		(14.2)	(8.7)	(11.4)
13	Others	0	1	1
		(0.0)	(0.2)	(0.1)
14	Non-response	21	22	43
		(3.4)	(3.5)	(3.5)
	Total	613	623	1236
		(100.0)	(100.0)	(100.0)

Note: Figures in parentheses refer to percentages to column totals.

was noticed in Lakhimpur district than in the other two; second, among these workers a substantially higher number of workers were found to be cultivators in Sivasagar district, while in Lakhimpur agricultural labour was the main occupation for a substantially large section of workers. Further, the share of non-agricultural work, both skilled and unskilled, as well as that of trade and business, though

TABLE 4.3
Occupational Distribution of Current Workers (Outside Tea Gardens)

Sl. No	Occupations	District Code			Total
		1	2	3	
1	Tea Garden Casual Labour	23	54	97	174
		(6.5)	(13.2)	(25.7)	(15.2)
2	Tea Garden Permanent Labour	4	11	4	19
		(1.1)	(2.7)	(1.1)	(1.7)
3	Tea Garden Lower Supervisory Staff & Service Providers	0	9	1	10
		(0.0)	(2.2)	(0.3)	(0.9)
4	Tea Garden Clerical Staff	2	1	0	3
		(0.6)	(0.2)	(0.0)	(0.3)
5	Cultivators	150	99	50	299
		(42.1)	(24.2)	(13.3)	(26.2)
6	Agricultural Labourers	10	18	114	142
		(2.8)	(4.4)	(30.2)	(12.4)
7	Non-farm Casual Labour (unskilled)	2	14	6	22
		(0.6)	(3.4)	(1.6)	(1.9)
8	Non-farm Casual Labour (skilled)	9	11	5	25
		(2.5)	(2.7)	(1.3)	(2.2)
9	Trade and Business	7	24	4	35
		(2.0)	(5.9)	(1.1)	(3.1)
10	Government Service	4	5	6	15
		(1.1)	(1.2)	(1.6)	(1.3)
11	Private Salaried Service	2	2	3	7
		(0.6)	(0.5)	(0.8)	(0.6)
12	Unemployed	136	147	71	354
		(38.2)	(35.9)	(18.8)	(31.0)
13	Others	1	0	2	3
		(0.3)	(0.0)	(0.5)	(0.3)
14	Non-response	6	14	14	34
		(1.7)	(3.4)	(3.7)	(3.0)
	Total	356	409	377	1142
		(100.0)	(100.0)	(100.0)	(100.0)

Note: Figures in parentheses refer to percentages to column totals.

small, was relatively higher in Dibrugarh than in the other two districts. Finally, the extent of reported unemployment was much higher in Sivasagar and Dibrugarh than in Lakhimpur. Behind these inter-district variations in the relative significance of different occupations lie the locational advantages and disadvantages faced by the workers.

So far as the gender differences in occupational distribution is concerned, it is noticed that female workers have a relatively higher share in the number of casual workers in the tea gardens, but have a lower share than the male workers in total number of permanent workers and other better paid workers in the tea gardens. Males have a greater presence in cultivation and agricultural labour, as well as in non-farm wage workers. A substantial proportion of female adults reported themselves as unemployed (Table 4.4). The significance of locational advantages and disadvantages seems to affect the occupational diversification of female workers as well. For example, while 41 per cent of female workers among the ex-tea labour households were found to be working in the tea gardens as casual labourers in Lakhimpur, 12 per cent of female workers in Sivasagar and 19.5 per cent in Dibrugarh could find casual work in the gardens. Again, in contrast to the other two districts, nearly 20 per cent of females worked as agricultural labourers in Lakhimpur. In the other two districts, non-agricultural employment was the main source of livelihoods for those who worked outside the tea gardens.

To sum up, the pattern of occupational diversification among adults belonging to tea garden worker households and ex-tea garden labour households brings out the relative significance of different livelihoods sources succinctly. First, the distinctions between 'tea garden labour households' and 'ex-tea garden labour households' seems to be of little analytical significance so far as the sources of livelihoods and employment are analysed. Among the tea garden labour households, a substantial portion of adults seek and get work in the tea gardens, but many of them, while living inside the tea gardens, depend on a variety of other sources for their livelihoods. Similarly, those who have left tea garden jobs, for a variety of reasons, have not completely severed their relation with the tea gardens. While the main breadwinners have moved out of the gardens to other occupations, other members of the households continue to depend on the gardens for their employment. Many of those who work in the gardens work as casual workers, mainly during peak plucking season. Second, there is a great deal of occupational concentration in the tea gardens. There is some occupational diversification among the labour households, but the overwhelming significance of tea gardens as a source of livelihoods can hardly be overstressed.

TABLE 4.4
Occupational Distribution of Male and Female Workers (Outside Tea Gardens)

Sl. No.	Occupations	Workers		
		Males	Females	All
1	Tea Garden Casual Labour	43	131	174
		(7.2)	(24.1)	(15.2)
2	Tea Garden Permanent Labour	14	5	19
		(2.3)	(0.9)	(1.7)
3	Tea Garden Lower Supervisory Staff & Service Providers	9	1	10
		(1.5)	(0.2)	(0.9)
4	Tea Garden Clerical Staff	3	0	3
		(0.5)	(0.0)	(0.3)
5	Cultivators	242	57	299
		(40.5)	(10.5)	(26.2)
6	Agricultural Labourers	94	48	142
		(15.7)	(8.8)	(12.4)
7	Non-farm Casual Labour (unskilled)	17	5	22
		(2.8)	(0.9)	(1.9)
8	Non-farm Casual Labour (skilled)	24	1	25
		(4.0)	(0.2)	(2.2)
9	Trade and Business	26	9	35
		(4.3)	(1.7)	(3.1)
10	Government Service	11	4	15
		(1.8)	(0.7)	(1.3)
11	Private Salaried Service	6	1	7
		(1.0)	(0.2)	(0.6)
12	Unemployed	99	255	354
		(16.6)	(46.9)	(31.0)
13	Others	2	1	3
		(0.3)	(0.2)	(0.3)
14	Non-response	8	26	34
		(1.3)	(4.8)	(3.0)
	Total	598	544	1142
		(100.0)	(100.0)	(100.0)

Note: Figures in parentheses refer to percentages to column totals.

The outside jobs are less in the nature of stable, permanent and better earning jobs like government service or private salaried service, most of it is within agriculture or casual work outside agriculture. Thirdly, within the tea gardens there seems to be a very insignificant presence of members of the labour households in the upper strata of jobs, such as those at the managerial, clerical and supervisory level.

Opportunities for vertical mobility within the gardens appear to be extremely limited.

Occupational Distribution of Fathers

In order to understand the extent and pattern of occupational shifts across generations, it would be useful to look at the occupational distribution of the fathers of the current workers. Table 4.5 presents the occupational distribution of fathers of the tea garden workers. It is found that, in the case of households categorised as living within the tea garden, more than 50 per cent of the fathers were engaged as permanent workers in the tea gardens. If we take into account all those who worked in the tea gardens, it comes to around 57 per cent. Unfortunately, a high percentage did not report their father's occupation. Leaving aside such cases of non-reporting, around 92 per cent of the fathers were found to be engaged within the tea gardens itself. There was hardly any significance of any other occupation in the previous generation. Another significant aspect is that this seems to be a consistent pattern in all the districts. In the case of those living outside the tea gardens, the extent of non-reporting is very high (see Table 4.6). If we ignore the cases of non-reporting, nearly 40 per cent of the fathers of the current workers outside the gardens were employed in the tea gardens — 33 per cent as permanent workers and 5.6 per cent as casual workers while the rest as lower supervisory staff and service providers. Among other occupations, of greater significance are cultivators and agricultural labourers. Together these two accounted for 40 per cent of workers, if the non-responses are ignored. Thus in the case of 'ex-tea garden' labourers households there seems to be some degree of diversification in the previous generation too. There also seems to be some difference in the districts in this regard as well.[3] In Lakhimpur district, ignoring the very high cases of non-reporting, nearly 70 per cent of the fathers were

---

[3] However, the data on occupation of fathers and grandfathers suffers from a number of inadequacies, which affects the reliability of the estimates. While the occupation of the father was meant to be the current occupation (provided the father is currently employed, or it was the major occupation for most part of his life), the answers were at times based on the recall of individuals, not always the most reliable.

TABLE 4.5
Occupational Distribution of Fathers (Within Tea Gardens)

Sl. No.	Occupations	District Code			Total
		1	*2*	*3*	
1	Tea Garden Casual Labour	8	0	20	28
		(2.3)	(0.0)	(4.5)	(2.3)
2	Tea Garden Permanent Labour	173	242	218	633
		(49.6)	(54.4)	(49.3)	(51.2)
3	Tea Garden Lower Supervisory Staff &	8	22	11	41
	Service Providers	(2.3)	(4.9)	(2.5)	(3.3)
4	Tea Garden Clerical Staff	0	3	0	3
		(0.0)	(0.7)	(0.0)	(0.2)
5	Cultivators	15	4	1	20
		(4.3)	(0.9)	(0.2)	(1.6)
6	Agricultural Labourers	7	1	2	10
		(2.0)	(0.2)	(0.5)	(0.8)
7	Non-farm Casual Labour (unskilled)	0	1	1	2
		(0.0)	(0.2)	(0.2)	(0.2)
8	Non-farm Casual Labour (skilled)	1	12	0	13
		(0.3)	(2.7)	(0.0)	(1.1)
9	Trade and Business	0	2	2	4
		(0.0)	(0.4)	(0.5)	(0.3)
10	Government Service	5	0	0	5
		(1.4)	(0.0)	(0.0)	(0.4)
12	Unemployed	0	2	2	4
		(0.0)	(0.4)	(0.5)	(0.3)
13	Others	0	1	0	1
		(0.0)	(0.2)	(0.0)	(0.1)
14	Non-response	132	155	185	472
		(37.8)	(34.8)	(41.9)	(38.2)
	Total	349	445	442	1236
		(100.0)	(100.0)	(100.0)	(100.0)

Note: Figures in parentheses refer to percentages to column totals.

engaged in the gardens, a substantial majority of them as permanent workers. In the cases of Sivasagar and Dibrugarh districts, the shares of those working in the tea gardens were 31.43 and 28.70 per cent respectively. Cultivators followed by agricultural labourers were the two prominent occupational groups outside the gardens in Sivasagar and Dibrugarh districts. In Lakhimpur, the relative share of agricultural labourers was higher than the other two districts.

TABLE 4.6
Occupational Distribution of Fathers (Outside Tea Gardens)

Sl. No.	Occupations	District Code 1	2	3	Total
1	Tea Garden Casual Labour	17	10	4	31
		(4.8)	(2.4)	(1.1)	(2.7)
2	Tea Garden Permanent Labour	37	48	95	180
		(10.4)	(11.7)	(25.2)	(15.8)
3	Tea Garden Lower Supervisory Staff & Service Providers	1	8	3	12
		(0.3)	(2.0)	(0.8)	(1.1)
4	Tea Garden Clerical Staff	4	3	0	7
		(1.1)	(0.7)	(0.0)	(0.6)
5	Cultivators	69	50	16	135
		(19.4)	(12.2)	(4.2)	(11.8)
6	Agricultural Labourers	23	39	22	84
		(6.5)	(9.5)	(5.8)	(7.4)
7	Non-farm Casual Labour (unskilled)	0	10	0	10
		(0.0)	(2.4)	(0.0)	(0.9)
8	Non-farm Casual Labour (skilled)	4	18	0	22
		(1.1)	(4.4)	(0.0)	(1.9)
9	Trade and Business	8	19	0	27
		(2.2)	(4.6)	(0.0)	(2.4)
10	Government Service	11	14	7	32
		(3.1)	(3.4)	(1.9)	(2.8)
11	Private Salaried Service	0	4	0	4
		(0.0)	(1.0)	(0.0)	(0.4)
12	Unemployed	0	3	0	3
		(0.0)	(0.7)	(0.0)	(0.3)
13	Others	1	4	0	5
		(0.3)	(1.0)	(0.0)	(0.4)
14	Non-response	181	179	230	590
		(50.8)	(43.8)	(61.0)	(51.7)
	Total	356	409	377	1142
		(100.0)	(100.0)	(100.0)	(100.0)

Note: Figures in parentheses refer to percentages to column totals.

Occupational Distribution of Grandfathers

Among the grandfathers of the current workers in the tea gardens, nearly 44 per cent were permanent tea garden workers, and there was virtually little diversification in that generation. If the non-responses are excluded, nearly 92 per cent of the grandfathers were in the tea gardens. Nearly 6.4 per cent were cultivators (Table 4.7). Among

TABLE 4.7
Occupational Distribution of Grandfathers (Within Tea Gardens)

Sl. No.	Occupations	District Code			Total
		1	*2*	*3*	*Total*
2	Tea Garden Permanent Labour	137	238	168	543
		(39.3)	(53.5)	(38.0)	(43.9)
5	Cultivators	26	10	2	38
		(7.4)	(2.2)	(0.5)	(3.1)
6	Agricultural Labourers	6	0	0	6
		(1.7)	(0.0)	(0.0)	(0.5)
7	Non-farm Casual Labour (unskilled)	0	2	0	2
		(0.0)	(0.4)	(0.0)	(0.2)
10	Government Service	4	0	0	4
		(1.1)	(0.0)	(0.0)	(0.3)
12	Unemployed	0	1	0	1
		(0.0)	(0.2)	(0.0)	(0.1)
14	Non-response	176	194	272	642
		(50.4)	(43.6)	(61.5)	(51.9)
	Total	349	445	442	1236
		(100.0)	(100.0)	(100.0)	(100.0)

Note: Figures in parentheses refer to percentages to column totals.

the grandfathers of the workers outside the gardens, 21.5 per cent were permanent tea garden workers. If the cases of non-response are excluded their share comes to nearly 50 per cent. Excluding such cases of non-reporting, nearly 33 per cent of the grandfathers of current workers outside the gardens were cultivators and 12 per cent were agricultural labourers (Table 4.8).

Intergenerational Occupational Mobility

As mentioned earlier, intergenerational occupational mobility can be analysed on the basis of transition matrices. Every element in these matrices represents the workers in different occupations as the proportion of the total number of sons/daughters of fathers in the same occupation i.

Thus, the elements of the intergenerational occupational matrix are defined as

$$a_{ij} = \frac{\sum O_{ij}}{\sum O_j}$$

TABLE 4.8
Occupational Distribution of Grandfathers (Outside Tea Gardens)

Sl. No.	Occupations	District Code 1	2	3	Total
2	Tea Garden Permanent Labour	57	97	92	246
		(16.0)	(23.7)	(24.4)	(21.5)
3	Tea Garden Lower Supervisory Staff	1	3	0	4
	& Service Providers	(0.3)	(0.7)	(0.0)	(0.4)
5	Cultivators	108	40	12	160
		(30.3)	(9.8)	(3.2)	(14.0)
6	Agricultural Labourers	10	39	10	59
		(2.8)	(9.5)	(2.7)	(5.2)
7	Non-farm Casual Labour (unskilled)	0	2	0	2
		(0.0)	(0.5)	(0.0)	(0.2)
9	Trade and Business	0	1	0	1
		(0.0)	(0.2)	(0.0)	(0.1)
10	Government Service	5	11	0	16
		(1.4)	(2.7)	(0.0)	(1.4)
14	Non-response	175	216	263	654
		(49.2)	(52.8)	(69.8)	(57.3)
	Total	356	409	377	1142
		(100.0)	(100.0)	(100.0)	(100.0)

Note: Figures in parentheses refer to percentages to column totals.

where, ΣO_{ij} is the number of workers in the i-th occupation whose fathers were in occupation j and ΣO_j is the total number of workers whose fathers were in occupation j.

The a_{ij} are typically interpreted as the outflow from fathers' occupation to son's occupation or transition probabilities from father's occupation to son's occupation. The diagonal element $a_{ij}:_i =_j$ represents the proportion of workers who have followed the same occupation as their fathers.

Thus, $1 - a_{ij},_i =_j$ can be taken as a measure of occupational mobility among the sons and daughters of the fathers who were in the occupation j.

The expression $\sum_{i \neq j} a_{ij}$, which is the proportion of workers working in an occupation other than their father's, is the extent of occupational mobility in the respective generation.

Occupational mobility of the tea garden can be seen at two levels: first, at the inter-sectoral level, that is, the extent to which tea garden labourers (or members of their households) have taken up an occupation outside the tea gardens; and secondly, within the tea sector, that is, the extent to which the workers (or their descendants) have moved from one level of job to another (notably higher) level of occupation.

Intergenerational Occupational Mobility: Current Workers

The intergenerational occupational mobility matrix[4] created on the basis of data generated from the field survey brings out several key aspects of the occupational change of workers in the tea garden. The data suffers from the following shortcomings. First, occupational transition, particularly across generations, should be seen in the wider context, that is, the occupational transition of one group must be seen in relation to the general levels of occupational transition in the economy. For obvious reasons of data constraints this is not possible in our case. The alternative that we have is the comparison across two groups of tea garden worker households. Second, there were substantial recall errors, particularly with respect to the occupation of the parents and grandparents. The recall lapses were more frequent in cases where the information about the occupations of parents and grandparents of female workers (wives and daughters-in-law) was sought from the male members of the households. Although attempts were made to recheck the information through subsequent visits, due to non-availability of the concerned members at the time of the visit by the investigators not all the information received was accurate. In order to partially overcome this lacuna, we have presented the transitional matrix without the non-response and including the

---

[4] The following notations have been used in the analysis to identify occupations: O_1: casual labour in tea garden; O_2: permanent worker in tea garden; O_3: lower supervisory staff and service providers in tea garden; O_4: clerical staff in tea gardens; O_5: cultivators; O_6: agricultural labourers; O_7: non-farm casual labour (unskilled); O_8: non-farm casual labour (skilled); O_9: trade and business; O_{10}: government service; O_{11}: private salaried service; O_{12}: unemployed; O_{13}: non-response.

non-responses. Further, we have presented the transition tables for male workers separately.

Tables 4.9 through 4.10 present the extent of intergenerational occupational mobility among workers in the tea garden worker households. All these workers belong to households still depending substantially on the tea garden for survival. As discussed earlier, the majority of workers of this category belong to two occupational groups: permanent labour and casual labour in tea gardens. Among workers whose fathers were tea garden labourers nearly 46 per cent have become permanent tea garden labourers, while among those whose fathers were casual labourers nearly 38 per cent are casual labourers in tea gardens. The distribution of fathers across occupations was found to be highly uneven, a majority of them working as labourers in the tea gardens, with very little occupational distribution in the earlier generation. Keeping that in mind, the results reported in these tables point out that among all the occupational categories reported here the highest intergenerational concentration (or immobility) of occupations has been in the categories of permanent and casual labourers. Among fathers engaged in other occupations, only 10 per cent of the sons and daughters of agricultural labourers and nearly 5 per cent of cultivators have followed their father's occupation. Nearly 75 per cent of workers whose fathers were cultivators have joined the tea gardens as permanent labourers. This trend is further strengthened if we ignore the cases where fathers' occupations have not been reported. Nearly 48 per cent of those working in the tea gardens as permanent labourers have their son and/or daughter in the same occupation. While nearly 40 per cent of workers whose fathers were casual labourers have followed their fathers' occupation. Thus, in comparison to other occupational categories, intergenerational occupational mobility was highest among tea garden labourers.

In Tables 4.11 and 4.12 we consider the cases of male workers alone, at first including the cases of non-response and then excluding such cases. A change that is noticed vis-à-vis the earlier table is that least occupational mobility is among the casual workers, followed by the permanent workers. However, the overall picture still suggests least occupational change across generations among the tea garden labourers. From Table 4.12, which corresponds to the occupational distribution of all male workers by the occupation of their fathers,

TABLE 4.9

Intergenerational Mobility Matrix: All Workers (Inside Tea Gardens)

Son's Occupation	O_1	O_2	O_3	O_4	O_5	O_6	O_7	O_8	O_9	O_{10}	O_{11}	O_{12}
												Father's Occupation
O_1	**0.464**	0.209	0.341	0	0.05	0.1	0	0.077	0	0	0	0.25
O_2	0.036	**0.382**	0.073	0	0.75	0.2	0.5	0	0	1	0	0.5
O_3	0	0.055	**0**	0	0	0.2	0	0	0.25	0	0	0
O_4	0	0.002	0	**0**	0	0	0	0	0	0	0	0
O_5	0.179	0.043	0.073	0	**0.05**	0.2	0	0	0	0	0	0
O_6	0	0.032	0	0	0	**0.1**	0	0	0	0	0	0
O_7	0	0.07	0.195	0	0	0.1	**0**	0.154	0.25	0	0	0
O_8	0	0.024	0	0	0	0	0	**0.077**	0	0	0	0
O_9	0	0.005	0	0.333	0.05	0	0	0	**0**	0	0	0
O_{10}	0	0.008	0	0.667	0.05	0	0	0	0	**0**	0	0
O_{11}	0	0.002	0	0	0	0	0	0	0	0	**0**	0
O_{12}	0.286	0.134	0.293	0	0.05	0.1	0.5	0.692	0.5	0	0	**0.25**

Note: Includes cases of non-reporting.

TABLE 4.10

Intergenerational Mobility Matrix: All Workers (Inside Tea Gardens)

Son's Occupation	Father's Occupation											
	O_1	O_2	O_3	O_4	O_5	O_6	O_7	O_8	O_9	O_{10}	O_{11}	O_{12}
O_1	**0.481**	0.216	0.350	0.000	0.050	0.100	0.000	0.077	0.000	0.000	0.000	0.250
O_2	0.037	**0.396**	0.075	0.000	0.750	0.200	0.500	0.000	0.000	1.000	0.000	0.500
O_3	0.000	0.057	**0.000**	0.000	0.000	0.200	0.000	0.000	0.250	0.000	0.000	0.000
O_4	0.000	0.002	0.000	**0.000**	0.000	0.000	0.000	0.000	0.000	0.000	0.000	0.000
O_5	0.185	0.044	0.075	0.000	**0.050**	0.200	0.000	0.000	0.000	0.000	0.000	0.000
O_6	0.000	0.033	0.000	0.000	0.000	**0.100**	0.000	0.000	0.000	0.000	0.000	0.000
O_7	0.000	0.072	0.200	0.000	0.050	0.100	**0.000**	0.000	0.250	0.000	0.000	0.000
O_8	0.000	0.025	0.000	0.000	0.000	0.000	0.000	**0.154**	0.000	0.000	0.000	0.000
O_9	0.000	0.005	0.000	0.333	0.000	0.000	0.000	0.077	**0.000**	0.000	0.000	0.000
O_{10}	0.000	0.008	0.000	0.667	0.050	0.000	0.000	0.000	0.000	**0.000**	0.000	0.000
O_{11}	0.000	0.002	0.000	0.000	0.000	0.000	0.000	0.000	0.000	0.000	**0.000**	0.000

Note: Cases of non-reporting of occupation have not been taken into consideration.

TABLE 4.11
Intergenerational Mobility Matrix: Male Workers (Inside Tea Gardens)

Son's Occupation	Father's Occupation											
	O_1	O_2	O_3	O_4	O_5	O_6	O_7	O_8	O_9	O_{10}	O_{12}	O_{13}
O_1	**0.286**	0.182	0.231	0	0.077	0.143	0	0	0	0	0	0.275
O_2	0.071	**0.427**	0.077	0	0.615	0.143	0.5	0	0	1	0.667	0.137
O_3	0	0.066	**0**	0	0	0.286	0	0	0.5	0	0	0.02
O_4	0	0.002	0	**0**	0	0	0	0	0	0	0	0.02
O_5	0.286	0.054	0.077	0	**0.077**	0.286	0	0	0	0	0	0.176
O_6	0	0.041	0	0	0	**0.143**	0	0	0	0	0	0.078
O_7	0	0.029	0.192	0	0.077	0	**0**	0.143	0	0	0	0.078
O_8	0	0.031	0	0	0	0	0	**0.143**	0	0	0	0.02
O_9	0	0.006	0	1	0.077	0	0	0	**0**	0	0	0.02
O_{10}	0	0.006	0	0	0.077	0	0	0	0	**0**	0	0.02
O_{12}	0.286	0.12	0.385	0	0.077	0	0.5	0.714	0.5	0	**0.333**	0.118
O_{13}	0.071	0.035	0.038	0	0	0	0	0	0	0	0	**0.039**

Note: Includes cases of non-reporting

TABLE 4.12

Intergenerational Mobility Matrix: Male Workers (Inside Tea Gardens)

Son's Occupation	Father's Occupation										
	O_1	O_2	O_3	O_4	O_5	O_6	O_7	O_8	O_9	O_{10}	O_{12}
O_1	**0.308**	0.189	0.240	0.000	0.077	0.143	0.000	0.000	0.000	0.000	0.000
O_2	0.077	**0.442**	0.080	0.000	0.615	0.143	0.500	0.000	0.000	1.000	0.667
O_3	0.000	0.069	**0.000**	0.000	0.000	0.286	0.000	0.000	0.500	0.000	0.000
O_4	0.000	0.002	0.000	**0.000**	0.000	0.000	0.000	0.000	0.000	0.000	0.000
O_5	0.308	0.056	0.080	0.000	**0.077**	0.286	0.000	0.000	0.000	0.000	0.000
O_6	0.000	0.043	0.000	0.000	0.000	**0.143**	0.000	0.000	0.000	0.000	0.000
O_7	0.000	0.030	0.200	0.000	0.077	0.000	**0.000**	0.000	0.000	0.000	0.000
O_8	0.000	0.032	0.000	0.000	0.000	0.000	0.000	**0.143**	0.000	0.000	0.000
O_9	0.000	0.006	0.000	1.000	0.000	0.000	0.000	0.000	**0.000**	0.000	0.000
O_{10}	0.000	0.006	0.000	0.000	0.077	0.000	0.000	0.000	0.000	**0.000**	0.000
O_{12}	0.308	0.124	0.400	0.000	0.077	0.000	0.500	0.714	0.500	0.000	**0.333**

Note: Cases of non-reporting of occupation have not been taken into consideration.

it is found that 24 per cent of the sons of those employed as tea garden supervisory staff and service providers became permanent tea garden workers. Thus there is still some concentration within the gardens, although vertically downward mobility[5] is noticed in this case. By and large there was very little vertically upward mobility within the tea gardens.

The picture of intergenerational occupational mobility is quite different in the cases of those who belong to the ex-tea garden labour households. In Table 4.13 it is reported that there is least occupational mobility among the cultivators. Nearly 55 per cent of sons and daughters of cultivator fathers have joined their father's occupation. While the percentages among the permanent tea labourers and casual tea labourers were 10 and 4 respectively. When we ignore the cases of non-reporting the relatively higher occupational immobility among the cultivators becomes even more pronounced (Table 4.14). When the cases of male workers alone are considered, it is found that 60 per cent of the sons of cultivators have followed their fathers occupations, while among the sons of agricultural labourers 16 per cent have done so. Nearly 13 per cent of the sons of unskilled casual labourers in the non-farm sector have followed their father's occupation, while among permanent and casual tea garden labourers their shares were 10 per cent and 4 per cent respectively (see Table 4.15). Ignoring the cases of non-reporting we find that broadly the picture of occupational mobility remains same (Table 4.16).

In order to bring out the differences in the intergenerational occupational mobility pattern of workers belonging to the tea labour households and ex-tea labour households, we have presented the percentage distribution of workers in different occupational categories, after

---

[5] Vertical mobility presupposes a hierarchical ordering of occupations. There are a number of ways to classify and order the different occupations either in terms of income or social status. One problem associated with such ordering is the *distance* among the ordered occupations. In our case, for example, casual labourers, permanent labourers, lower supervisory staff and managerial staff can be thought of as an ascending order of occupations within the gardens, the distance between casual and permanent labourers is not the same as that between permanent labourers and lower supervisory staff or that between the later and managerial staff. For the purpose of our analysis, we have not attempted to quantify the distance between the occupations.

TABLE 4.13

Intergenerational Mobility Matrix: All Workers (Outside Tea Gardens)

Son's Occupation	Father's Occupation												
	O_1	O_2	O_3	O_4	O_5	O_6	O_7	O_8	O_9	O_{10}	O_{11}	O_{12}	O_{13}
O_1	**0.097**	0.106	0.083	0.000	0.022	0.131	0.000	0.136	0.185	0.188	0.500	0.000	0.203
O_2	0.065	**0.039**	0.000	0.000	0.022	0.000	0.000	0.045	0.000	0.031	0.000	0.000	0.008
O_3	0.000	0.000	**0.000**	0.000	0.000	0.036	0.000	0.000	0.074	0.000	0.000	0.667	0.005
O_4	0.000	0.011	0.000	**0.000**	0.007	0.000	0.000	0.000	0.000	0.000	0.000	0.000	0.000
O_5	0.226	0.350	0.333	0.000	**0.548**	0.429	0.300	0.091	0.111	0.250	0.000	0.000	0.166
O_6	0.032	0.256	0.000	0.000	0.037	**0.131**	0.200	0.045	0.000	0.000	0.000	0.000	0.128
O_7	0.065	0.033	0.000	0.000	0.015	0.024	**0.100**	0.000	0.000	0.031	0.000	0.000	0.013
O_8	0.065	0.067	0.083	0.000	0.022	0.000	0.200	**0.000**	0.000	0.031	0.000	0.000	0.007
O_9	0.097	0.033	0.000	0.000	0.044	0.024	0.000	0.000	**0.037**	0.000	0.000	0.000	0.029
O_{10}	0.000	0.011	0.000	0.143	0.030	0.000	0.100	0.000	0.037	**0.031**	0.250	0.000	0.007
O_{11}	0.000	0.000	0.000	0.000	0.000	0.000	0.000	0.045	0.000	0.031	**0.250**	0.000	0.007
O_{12}	0.323	0.072	0.500	0.857	0.230	0.190	0.100	0.545	0.556	0.375	0.000	**0.333**	0.388
O_{13}	0.032	0.022	0.000	0.000	0.022	0.036	0.000	0.091	0.000	0.031	0.000	0.000	**0.039**

Note: Includes cases of non-reporting.

TABLE 4.14

Intergenerational Mobility Matrix: All Workers (Outside Tea Gardens)

Son's Occupation	Father's Occupation											
	O_1	O_2	O_3	O_4	O_5	O_6	O_7	O_8	O_9	O_{10}	O_{11}	O_{12}
O_1	**0.100**	0.108	0.083	0.000	0.023	0.136	0.000	0.150	0.185	0.194	0.500	0.000
O_2	0.067	**0.040**	0.000	0.000	0.023	0.000	0.000	0.050	0.000	0.032	0.000	0.000
O_3	0.000	0.000	**0.000**	0.000	0.000	0.037	0.000	0.000	0.074	0.000	0.000	0.667
O_4	0.000	0.011	0.000	**0.000**	0.008	0.000	0.000	0.000	0.000	0.000	0.000	0.000
O_5	0.233	0.358	0.333	0.000	**0.561**	0.444	0.300	0.100	0.111	0.258	0.000	0.000
O_6	0.033	0.261	0.000	0.000	0.038	**0.136**	0.200	0.050	0.000	0.000	0.000	0.000
O_7	0.067	0.034	0.000	0.000	0.015	0.025	**0.100**	0.000	0.000	0.032	0.000	0.000
O_8	0.067	0.068	0.083	0.000	0.023	0.000	0.200	**0.000**	0.000	0.032	0.000	0.000
O_9	0.100	0.034	0.000	0.000	0.045	0.025	0.000	0.000	**0.037**	0.000	0.000	0.000
O_{10}	0.000	0.011	0.000	0.143	0.030	0.000	0.100	0.000	0.037	**0.032**	0.250	0.000
O_{11}	0.000	0.000	0.000	0.000	0.000	0.000	0.000	0.050	0.000	0.032	**0.250**	0.000
O_{12}	0.333	0.074	0.500	0.857	0.235	0.198	0.100	0.600	0.556	0.387	0.000	**0.333**

Note: Cases of non-reporting of occupation have not been taken into consideration.

TABLE 4.15

Intergenerational Mobility Matrix: Male Workers (Outside Tea Gardens)

Son's Occupation	Father's Occupation												
	O_1	O_2	O_3	O_4	O_5	O_6	O_7	O_8	O_9	O_{10}	O_{11}	O_{12}	O_{13}
O_1	**0.100**	0.089	0.000	0.000	0.017	0.031	0.000	0.182	0.118	0.222	0.000	0.000	0.088
O_2	0.100	**0.038**	0.000	0.000	0.026	0.000	0.000	0.000	0.000	0.056	0.000	0.000	0.012
O_3	0.000	0.000	**0.000**	0.000	0.000	0.047	0.000	0.000	0.118	0.000	0.000	1.000	0.012
O_4	0.000	0.013	0.000	**0.000**	0.009	0.000	0.000	0.000	0.000	0.000	0.000	0.000	0.000
O_5	0.200	0.389	0.400	0.000	**0.600**	0.531	0.375	0.000	0.176	0.278	0.000	0.000	0.347
O_6	0.000	0.261	0.000	0.000	0.043	**0.156**	0.250	0.091	0.000	0.000	0.000	0.000	0.206
O_7	0.100	0.038	0.000	0.000	0.017	0.031	**0.125**	0.000	0.000	0.056	0.000	0.000	0.018
O_8	0.100	0.076	0.100	0.000	0.017	0.000	0.250	**0.000**	0.000	0.056	0.000	0.000	0.024
O_9	0.150	0.032	0.000	0.000	0.052	0.031	0.000	0.000	**0.059**	0.000	0.000	0.000	0.053
O_{10}	0.000	0.013	0.000	0.200	0.035	0.000	0.000	0.000	0.059	**0.056**	0.000	0.000	0.012
O_{11}	0.000	0.000	0.000	0.000	0.000	0.000	0.000	0.091	0.000	0.000	**1.000**	0.000	0.024
O_{12}	0.250	0.025	0.500	0.800	0.165	0.156	0.000	0.545	0.471	0.222	0.000	**0.000**	0.200
O_{13}	0.000	0.025	0.000	0.000	0.017	0.016	0.000	0.091	0.000	0.056	0.000	0.000	**0.006**

Note: Includes cases of non-reporting.

TABLE 4.16

Intergenerational Mobility Matrix: Male Workers (Outside Tea Gardens)

Son's Occupation	Father's Occupation											
	O_1	O_2	O_3	O_4	O_5	O_6	O_7	O_8	O_9	O_{10}	O_{11}	O_{12}
O_1	**0.100**	0.092	0.000	0.000	0.018	0.032	0.000	0.200	0.118	0.235	0.000	0.000
O_2	0.100	**0.039**	0.000	0.000	0.027	0.000	0.000	0.000	0.000	0.059	0.000	0.000
O_3	0.000	0.000	**0.000**	0.000	0.000	0.048	0.000	0.000	0.118	0.000	0.000	1.000
O_4	0.000	0.013	0.000	**0.000**	0.009	0.000	0.000	0.000	0.000	0.000	0.000	0.000
O_5	0.200	0.399	0.400	0.000	**0.611**	0.540	0.375	0.000	0.176	0.294	0.000	0.000
O_6	0.000	0.268	0.000	0.000	0.044	**0.159**	0.250	0.100	0.000	0.000	0.000	0.000
O_7	0.100	0.039	0.000	0.000	0.018	0.032	**0.125**	0.000	0.000	0.059	0.000	0.000
O_8	0.100	0.078	0.100	0.000	0.018	0.000	0.250	**0.000**	0.000	0.059	0.000	0.000
O_9	0.150	0.033	0.000	0.000	0.253	0.032	0.000	0.000	**0.059**	0.000	0.000	0.000
O_{10}	0.000	0.013	0.000	0.200	0.335	0.000	0.000	0.000	0.059	**0.059**	0.000	0.000
O_{11}	0.000	0.000	0.000	0.000	0.000	0.000	0.000	0.100	0.000	0.000	**1.000**	0.000
O_{12}	0.250	0.026	0.500	0.800	0.168	0.159	0.000	0.600	0.471	0.235	0.000	**0.000**

Note: Cases of non-reporting of occupation have not been taken into consideration.

dividing them into two categories, viz. those whose fathers were in some occupations in the tea gardens and those who were not (see Tables 4.17 and 4.18). It is found that among the tea labour households (living within the tea gardens), nearly 65 per cent of those whose fathers were working in the gardens have found a job within the gardens, while 54 per cent of those whose fathers were not in the gardens also work in the gardens. Among those who belong to 'ex-tea garden labour households' and live outside the gardens, the scenario is somewhat different. Nearly 15 per cent of those whose fathers worked in the gardens were employed in the gardens themselves, while 18 per cent of those whose fathers were working outside the gardens joined the gardens. A majority of the workers of course were engaged either as cultivators or as agricultural labourers.

Intergenerational Occupational Mobility: Earlier Generations

The intergenerational occupational mobility of earlier generations is analysed to provide a backdrop to the ongoing process of occupational change among the tea garden workers in upper Assam.

TABLE 4.17
Distribution of Workers in Occupations According to Father's Occupation (Within Gardens)

Son's Occupation	All Workers		Male Workers	
	Fathers in Tea Gardens	*Fathers in Other Occupations*	*Fathers in Tea Gardens*	*Fathers in Other Occupations*
O_1	23.35	6.78	18.70	17.98
O_2	36.12	44.07	39.89	25.84
O_3	5.14	5.08	6.11	4.49
O_4	0.15	0.00	0.19	1.12
O_5	5.14	5.08	6.11	13.48
O_6	2.94	1.69	3.82	5.62
O_7	7.64	6.78	3.63	5.62
O_8	2.20	3.39	2.86	3.37
O_9	0.59	0.00	0.76	1.12
O_{10}	1.03	1.69	0.57	2.25
O_{11}	0.15	0.00	0.00	0.00
O_{12}	15.42	25.42	13.74	16.85
O_{13}	0.15	0.00	3.63	2.25

TABLE 4.18
Distribution of Workers in Occupations According to Father's Occupation
(Outside Gardens)

	All Workers		Male Workers	
Son's Occupation	Fathers in Tea Gardens	Fathers in Other Occupations	Fathers in Tea Gardens	Fathers in Other Occupations
O_1	10.00	16.56	8.33	6.65
O_2	3.91	1.10	4.17	1.48
O_3	0.00	1.10	0.00	2.22
O_4	0.87	0.11	1.04	0.25
O_5	32.17	24.67	35.94	42.61
O_6	20.43	10.42	21.35	13.05
O_7	3.48	1.54	4.17	2.22
O_8	6.52	1.10	7.81	2.22
O_9	3.91	2.85	4.17	4.43
O_{10}	1.30	1.32	1.56	1.97
O_{11}	0.00	0.77	0.00	1.48
O_{12}	15.22	34.98	9.38	19.95
O_{13}	2.17	3.51	2.08	1.48

Thus we have constructed the intergenerational occupational matrix of the earlier generation by cross classifying the occupation of the fathers of the current workers with that of the grandfathers. This exercise, not withstanding its serious limitations on account of the non-reporting as well as recall errors in reporting of data regarding earlier generations of workers, still provides a reliable basis to understand intergenerational occupational changes in the population under study. As discussed earlier, we have constructed the marginal ratios of occupational transition with respect to two distinct sets of workers, viz. those belonging to tea labour households and those who belong to ex-tea labour households. It is found that in the earlier generation, the extent of occupational change was considerably less pronounced than that in the current generations. For example, among the tea garden households, 80 per cent of those whose fathers were permanent labourers, themselves became permanent labourers in the gardens (Table 4.19), which increases to 82 per cent if we drop the cases of non-response (Table 4.20). Similarly among the ex-tea garden labour households living outside the gardens, 47 per cent of those whose fathers were cultivators became cultivators (Table 4.21). If the cases of non-reporting are disregarded, the share goes up to 65 per cent (Table 4.22).

TABLE 4.19
Intergenerational Mobility Matrix (Earlier Generation): All Workers
(Inside Tea Gardens)

(Including non-response)

Father's Occupation	Grandfather's Occupation						
	O_2	O_5	O_6	O_7	O_{10}	O_{12}	O_{13}
O_1	0.041	0.026	0.000	0.000	0.250	0.000	0.006
O_2	0.807	0.632	0.000	1.000	0.500	1.000	0.259
O_3	0.076	0.000	0.000	0.000	0.000	0.000	0.000
O_4	0.006	0.000	0.000	0.000	0.000	0.000	0.000
O_5	0.011	0.263	0.167	0.000	0.000	0.000	0.005
O_6	0.006	0.000	0.833	0.000	0.000	0.000	0.003
O_7	0.004	0.000	0.000	0.000	0.000	0.000	0.000
O_8	0.022	0.026	0.000	0.000	0.000	0.000	0.000
O_9	0.004	0.053	0.000	0.000	0.000	0.000	0.000
O_{10}	0.007	0.000	0.000	0.000	0.250	0.000	0.000
O_{12}	0.007	0.000	0.000	0.000	0.000	0.000	0.000
O_{13}	0.011	0.000	0.000	0.000	0.000	0.000	0.727

TABLE 4.20
Intergenerational Mobility Matrix (Earlier Generation): All Workers
(Inside Tea Gardens)

(Excluding non-response)

Father's Occupation	Grandfather's Occupation					
	O_2	O_5	O_6	O_7	O_{10}	O_{12}
O_1	0.041	0.026	0.000	0.000	0.250	0.000
O_2	0.816	0.632	0.000	1.000	0.500	1.000
O_3	0.076	0.000	0.000	0.000	0.000	0.000
O_4	0.006	0.000	0.000	0.000	0.000	0.000
O_5	0.011	0.263	0.167	0.000	0.000	0.000
O_6	0.006	0.000	0.833	0.000	0.000	0.000
O_7	0.004	0.000	0.000	0.000	0.000	0.000
O_8	0.022	0.026	0.000	0.000	0.000	0.000
O_9	0.004	0.053	0.000	0.000	0.000	0.000
O_{10}	0.007	0.000	0.000	0.000	0.250	0.000
O_{12}	0.007	0.000	0.000	0.000	0.000	0.000

TABLE 4.21
Intergenerational Mobility Matrix (Earlier Generation): All Workers
(Outside Tea Gardens)

(Including non-response)

Father's Occupation	Grandfather's Occupation							
	O_2	O_3	O_5	O_6	O_7	O_9	O_{10}	O_{13}
O_1	0.069	0.000	0.050	0.068	0.000	0.000	0.000	0.003
O_2	0.378	0.000	0.019	0.085	0.000	0.000	0.063	0.119
O_3	0.024	0.000	0.000	0.000	0.000	0.000	0.063	0.008
O_4	0.020	0.000	0.013	0.000	0.000	0.000	0.000	0.000
O_5	0.065	0.000	0.469	0.017	0.000	0.000	0.313	0.058
O_6	0.114	0.000	0.069	0.254	0.000	1.000	0.000	0.044
O_7	0.000	0.750	0.000	0.119	0.000	0.000	0.000	0.000
O_8	0.077	0.000	0.000	0.000	1.000	0.000	0.000	0.002
O_9	0.012	0.000	0.038	0.136	0.000	0.000	0.375	0.006
O_{10}	0.024	0.000	0.063	0.000	0.000	0.000	0.188	0.020
O_{11}	0.000	0.000	0.000	0.000	0.000	0.000	0.000	0.006
O_{12}	0.000	0.000	0.000	0.051	0.000	0.000	0.000	0.000
O_{13}	0.215	0.250	0.281	0.271	0.000	0.000	0.000	0.734

TABLE 4.22
Intergenerational Mobility Matrix (Earlier Generation): All Workers
(Outside Tea Gardens)

(Excluding non-response)

Father's Occupation	Grandfather's Occupation						
	O_2	O_3	O_5	O_6	O_7	O_9	O_{10}
O_1	0.088	0.000	0.070	0.093	0.000	0.000	0.000
O_2	0.482	0.000	0.026	0.116	0.000	0.000	0.063
O_3	0.031	0.000	0.000	0.000	0.000	0.000	0.063
O_4	0.026	0.000	0.017	0.000	0.000	0.000	0.000
O_5	0.083	0.000	0.652	0.023	0.000	0.000	0.313
O_6	0.145	0.000	0.096	0.349	0.000	1.000	0.000
O_7	0.000	1.000	0.000	0.163	0.000	0.000	0.000
O_8	0.098	0.000	0.000	0.000	1.000	0.000	0.000
O_9	0.016	0.000	0.052	0.186	0.000	0.000	0.375
O_{10}	0.031	0.000	0.087	0.000	0.000	0.000	0.188
O_{11}	0.000	0.000	0.000	0.000	0.000	0.000	0.000
O_{12}	0.000	0.000	0.000	0.070	0.000	0.000	0.000

However, there was significant intergenerational mobility between two occupations — tea garden labour and cultivation. It was found that, in the case of tea labour households, 63 per cent of those whose fathers were cultivators joined the tea gardens as casual labourers. Similarly, among the ex-tea garden labour households outside the gardens, 7–10 per cent of those whose fathers were cultivators joined the gardens. Thus it is clear that the dependence upon tea gardens was quite significant for the earlier generations. The extent of occupational mobility appears to be very low. Most of those whose fathers were tea garden labourers joined the gardens as labourers and most of those whose fathers were cultivators continued to pursue the same occupation. There was very little vertical movement in terms of changes to better-paid occupations within the gardens.

Determinants of Occupational Diversification and Mobility

In order to find out the factors influencing the movement of tea garden labour to the labour market, binary logistic regression was estimated. The occupational categorisation of workers as tea and non-tea labour in this case needs some clarification. We have two categories of households; first, those who are categorised as 'tea labour households within the gardens'. In the first category of households the main source of livelihoods is the tea garden and some members move outside the garden for employment. The labour that is categorised as 'non-tea garden labour' here are those who belong to the tea labour *households*, but have occupations other than those in the tea gardens. In the second case, the households do not live inside the gardens; they mostly depend upon sources other than the tea garden for their livelihoods. However, some members of such 'outside tea garden' households also work inside the gardens. In either case we try to find out the determinants of participation in non-tea labour occupations on the part of the labourers. The exercise is limited to the workers belonging to the current generation only.

To investigate the determinants of participation in non-tea occupations, we have estimated a binary logistic model where the dependent variable takes the value 1, when the individual has access to an occupation outside the garden and 0 otherwise. The results for binary logistic regression to find out the determinants of occupations other than those in the tea gardens are presented in Tables 4.23 and 4.24. The variable specifications are as follows.

Dependent Variable:	
Z_i	1 if the individual has an occupation outside the tea garden, otherwise = 0
Independent Variables:	
AGE	Age of the individual in years
AGE SQ	Square of the age of the individual in years
GENDER	Male = 1, Female = 0
FATHEROCC	Occupation of the father of the individual = 1 if works outside the tea garden, otherwise = 0
EDUCAT	Education dummy = 1, if completed 8 years of schooling, otherwise 0
SKILL	Skill dummy = 1, if the individual is a skilled labour, otherwise = 0
URBAN	Nearness to urban location dummy = 1 if located near an urban locality, otherwise = 0

It is found that among the tea labour households the significant determinants of occupational diversification are age, gender, educational qualification of the worker and the nearness of the garden to an urban location. While a relatively higher age decreases the probability of entering a non-tea labour occupation, being female increases the chances of being in the non-tea labour occupation. Education up to eight years of schooling increases the possibility of being absorbed outside the tea garden significantly. Nearness to an urban location has a positive impact on occupational diversification at the individual level. The findings are broadly consistent with the results discussed earlier. The households within tea gardens move out only when some members have higher educational status or when, as a result of non-availability of employment within the gardens, some members, particularly the younger and female members, go out to find alternative employment opportunities outside the gardens. At the household level, the strategy seems to be to augment the family earnings without moving out of the tea garden.

On the other hand, the determinants of occupational diversification among individuals in the households living outside the gardens seem to be somewhat different. The significant variables are age, gender and nearness to urban location. While with higher age the probability of entering non-tea occupation increases, being a male also increases

TABLE 4.23
Determinants of Occupational Diversification: Inside the Gardens

Variables	B	S.E.	Exp (B)
AGE	−0.122[*]	0.026	0.885
AGE SQ	0.001[*]	0.000	1.001
GENDER	−0.484[*]	0.150	0.617
FATHEROCC	0.034	0.378	1.034
EDUCAT	0.871[*]	0.183	2.389
URBAN	0.350	0.157	1.419
CONSTANT	1.290	0.455	3.633
−2 LOGLIKELIHOOD	1180.628		

Note: [*]indicates significance at 0.001 level.

TABLE 4.24
Determinants of Occupational Diversification: Outside the Gardens

Variables	B	S.E.	Exp (B)
AGE	0.137[*]	0.032	1.147
AGE SQ	−0.002[*]	0.000	0.998
GENDER	1.834[*]	0.197	6.259
FATHEROCC	0.150	0.235	1.162
EDUCAT	−0.209	0.243	0.811
SKILL	1.350	0.775	3.856
URBAN	1.068[*]	0.238	2.909
CONSTANT	−2.903	0.615	0.055
−2 LOGLIKELIHOOD	711.950		

Note: [*]indicates significance at 0.001 level.

the probability of working in a non-tea occupation. These results are contradictory to the earlier findings. Nearness to urban areas is positively associated with the diversification process. The reasons of these apparently contradictory findings could be due to the different kinds of constraints faced by the households who have already moved out of the tea gardens. Here, the mainstay remains the occupations outside the tea gardens, but female workers typically join the tea garden occupation during the peak plucking seasons. Since many of the workers work in agriculture or the low skill end of the non-farm market, the level of education does not significantly affect their occupational movement to non-tea garden occupations.

Conclusion

The pattern of occupational diversification among adults belonging to tea garden worker households and ex-tea garden labour households brings out the relative significance of different livelihoods sources succinctly. First, the distinctions between 'tea garden labour households' and 'ex-tea garden labour households' seems to be of little analytical significance so far as the sources of livelihood and employment are analysed. Among the tea garden labour households, a substantial portion of adults seek and get work in the tea gardens, but many of them, while living inside the tea gardens, depend on a variety of other sources for their livelihood. Similarly, those who have left the tea garden jobs, for a variety of reasons, have not completely severed their relation with the tea gardens. While the main breadwinners have moved out of the gardens to other occupations, either out of compulsion or choice, other members of the households continue to depend on the gardens for their employment (Mishra et al. 2011). Many of those who work in the gardens work as casual workers, mainly in the peak plucking season. Second, there is a great deal of occupational concentration in the tea gardens. There is some occupational diversification among the labour households, but the overwhelming significance of tea gardens as a source of livelihoods can hardly be overstressed. The outside jobs are less in the nature of stable, permanent and better earning jobs like government service or private salaried service, most of it is within agriculture or casual work outside agriculture. Third, within the tea gardens there seems to be a very insignificant presence of members of the labour households in the upper strata of jobs, such as those at the managerial, clerical and supervisory level. Opportunities for vertical mobility within the gardens appear to be extremely limited. Thus, in the previous generation tea gardens were the main source of employment and there was hardly any occupational diversification among those who lived in the gardens. Another significant aspect is that this seems to be a consistent pattern in all the districts.

The intergenerational occupational mobility matrix created on the basis of data generated from the field survey brings out several key aspects of the occupational change of workers in the tea garden.

Among all the occupational categories reported here the highest intergenerational concentration (or immobility) of occupations has been in the categories of permanent and casual labourers. It is found that among the tea labour households (living within the tea gardens), nearly 65 per cent of those whose fathers were working in the gardens have found a job within the gardens, while 54 per cent of those whose fathers were not in the gardens also work in the gardens. Among those who belong to 'ex-tea garden labour households' and live outside the gardens, the scenario is somewhat different. Nearly 15 per cent of those whose fathers had worked in the gardens were employed in the gardens themselves, while 18 per cent of those whose fathers were working outside gardens have joined the gardens. A majority of workers of course were engaged either as cultivators or as agricultural labourers.

It is found that among the tea labour households the significant determinants of occupational diversification are age, gender, and educational qualification of the worker and the nearness of the garden to an urban location. While a relatively higher age decreases the probability of entering non-tea labour occupation, being female increases the chances of being in the non-tea labour occupation. Education up to eight years of schooling increases the possibility of being absorbed outside the tea garden significantly. Nearness to urban location has a positive impact on occupational diversification at the individual level. On the other hand the determinants of occupational diversification among the individuals in the households living outside the gardens seem to be somewhat different. The significant variables are age, gender and nearness to urban location. While with higher age the probability of entering non-tea occupation increases, being a male also increases the probability of working in a non-tea occupation.

5

Causes and Implications of Occupational Mobility

Shifting of labour, from one occupation to another, is a result of both push and pull factors of various kinds. The relative importance of the various factors is difficult to ascertain because of the simultaneous existence of various forces that prompt the individual or the household for the occupational shift. However, while there is considerable evidence in the literature that expected or actual levels of income play a prominent role in the decision making about occupational change, there are a host of other variables which influence an individual's choices. We attempted to identify a few factors that influence occupational diversification in the previous chapter. In this chapter we carry forward the analysis, by focusing on the perception of the workers with regard to occupational change and diversification. This discussion is supplemented by a discussion of the factors encouraging or discouraging the workers to move out of the tea sector.

Perceptions on Working Conditions and Occupational Change

Workers' perception on their working conditions and the desire to change occupations provide important insights regarding the conditions of work inside and outside the gardens. However, an analysis of perceptions through a questionnaire method has some inherent limitations.[1] Respondents may not be willing to share

---

[1] The perceptions, for example, could be influenced by the way the questions were posed and asked, the power relations between the interviewer and the interviewee, the gaps in the understanding between them, and also the situation in which the interview is conducted.

their real perceptions with the interviewer for a number of reasons. Moreover, such revealed perceptions recorded with reference to hypothetical situations may not reflect the real choices opted for in concrete situations. Although all possible steps were taken to minimise the errors and misreporting on these counts, still the results are, to some extent, influenced by these factors.

Among all categories of workers within the gardens, dissatisfaction with working conditions was highest among the casual tea garden workers and those engaged in trade and business. Over all, an overwhelmingly large number of workers have reported to be satisfied with their working conditions (Table 5.1). Outside, the garden, workers were found to be more ambiguous and less satisfied with their working conditions. Those who were found to be most

TABLE 5.1
Worker's Satisfaction in Current Occupation: Tea Garden

Code No.	Occupation	Satisfied with Present Working Condition			
		Not Satisfied	Satisfied	No Opinion	Total
1	Cultivators	0	13	0	13
		(0.00)	(100.0)	(0.00)	(100.0)
2	Agricultural Labourers	0	3	0	3
		(0.00)	(100.0)	(0.00)	(100.0)
3	Tea garden Casual Labourer	14	20	0	34
		(41.2)	(58.8)	(0.00)	(100.0)
4	Tea Garden Permanent Labourer	8	286	3	297
		(2.7)	(96.3)	(1.0)	(100.0)
5	Tea Garden Supervisory Staff	0	7	0	7
		(0.00)	(100.0)	(0.00)	(100.0)
8	Non-Farm Wage Labour	1	6	0	7
		(14.3)	(85.7)	(0.00)	(100.0)
10	Trade and Business	2	0	0	2
		(100.0)	(0.00)	(0.00)	(100.0)
14	Not Employed	1	4	0	5
		(20.0)	(80.0)	(0.00)	(100.0)
15	Others	0	7	0	7
		(0.00)	(100.0)	(0.00)	(100.0)
	Total	26	346	3	375
		(6.9)	(92.3)	(0.8)	(100.0)

Note: Figures in the parentheses refer to percentage to row totals.
Source: Field survey (Tables 5.1–5.20)

dissatisfied with their working conditions include private salaried employees, non-farm casual worker and agricultural labourers (Table 5.2). Interestingly, casual tea garden workers outside the gardens are less dissatisfied about their working conditions than their counterparts inside the gardens. Among those who stay outside the gardens, nearly 4 per cent have reported that the shift to their own occupation was not voluntary (Table 5.3). To delve further into the perceptions of those who live outside the garden, their views with respect to the tea gardens have been presented in Table 5.4. Not many among those who have moved out of the tea gardens would

TABLE 5.2
Perceptions Regarding Working Conditions: Outside Tea Garden

Code No.	Occupations	*Satisfied With Working Condition*			
		No	*Yes*	*No Opinion*	*Total*
1	Cultivators	13	53	49	115
		(11.3)	(46.1)	(42.6)	(100.0)
2	Agricultural Labourers	17	13	55	85
		(20.0)	(15.3)	(64.7)	(100.0)
3	Tea garden Casual Labourer	2	9	4	15
		(13.3)	(60.0)	(26.7)	(100.0)
4	Tea Garden Permanent Labourer	1	5	1	7
		(14.3)	(71.4)	(14.3)	(100.0)
5	Tea Garden Supervisory Staff	0	1	0	1
		(0.00)	(100.0)	(0.00)	(100.0)
8	Non-Farm Wage Labour	17	48	4	69
		(24.6)	(69.6)	(5.8)	(100.0)
10	Trade and Business	1	18	15	34
		(2.9)	(52.9)	(44.1)	(100.0)
11	Government Service	1	8	6	15
		(6.7)	(53.3)	(40.0)	(100.0)
12	Private Salaries Job	2	2	0	4
		(50.0)	(50.0)	(0.00)	(100.0)
14	Not Employed	2	3	4	9
		(22.2)	(33.3)	(44.4)	(100.0)
15	Others	3	14	3	20
		(15.0)	(70.0)	(15.0)	(100.0)
	Total	59	174	141	374
		(15.8)	(46.5)	(37.7)	(100.0)

Note: Figures in the parentheses refer to percentage to row totals.

TABLE 5.3
Shift of Occupation Voluntary or Not: Outside Tea Garden

Code No.	Occupations	*Was Shift to Present Occupation Voluntary*				
		No	Yes	Can't Say	Not Applicable	Total
1	Cultivators	6	1	41	67	115
		(5.2)	(0.9)	(35.7)	(58.3)	(100.0)
2	Agricultural Labourers	3	0	46	36	85
		(3.5)	(0.00)	(54.1)	(42.4)	(100.0)
3	Tea garden Casual Labourer	1	2	3	9	15
		(6.7)	(3.4)	(20.0)	(60.0)	(100.0)
4	Tea Garden Permanent Labourer	0	1	4	2	7
		(0.00)	(14.3)	(57.1)	(28.6)	(100.0)
5	Tea Garden Supervisory Staff	0	0	1	0	1
		(0.00)	(0.00)	(100.0)	(0.00)	(100.0)
8	Non-Farm Wage Labour	1	2	4	62	69
		(1.4)	(2.9)	(5.8)	(89.9)	(100.0)
10	Trade and Business	0	1	12	21	34
		(0.00)	(2.9)	(35.3)	(61.8)	(100.0)
11	Government Service	0	2	7	6	15
		(0.00)	(13.3)	(46.7)	(40.0)	(100.0)
12	Private Salaries Job	1	0	0	3	4
		(25.0)	(0.00)	(0.00)	(75.0)	(100.0)
14	Not Employed	1	1	4	3	9
		(11.1)	(11.1)	(44.4)	(33.3)	(100.0)
15	Others	2	1	3	14	20
		(10.0)	(5.0)	(15.0)	(70.0)	(100.0)
	Total	15	11	125	223	374
		(4.0)	(3.0)	(33.4)	(59.6)	(100.0)

Note: Figures in the parentheses refer to percentage to row totals.

like to return to the tea garden job. The disenchantment with tea garden jobs was sharper in the case of those who had salaried jobs in government or private sectors, and those working as non-farm labour. Although not much can be read into the findings of this nature, it does indicate the complexities associated with the perceptions of individuals regarding their occupations.

Within the gardens, the desire to change the current employer is remarkably high among the casual tea garden workers although in total very few workers have expressed their opinion in favour of

TABLE 5.4
Perceptions Regarding Tea Garden Jobs: Outside Tea Garden

Code No.	Occupations	Like to Work in a Tea Garden Once Again			
		No	Yes	Can't Say	Total
1	Cultivators	39	3	73	115
		(33.9)	(2.6)	(63.5)	(100.0)
2	Agricultural	20	3	62	85
	Labourers	(23.5)	(3.5)	(72.9)	(100.0)
3	Tea garden Casual	8	1	6	15
	Labourer	(53.3)	(6.7)	(40.0)	(100.0)
4	Tea Garden	1	1	5	7
	Permanent Labourer	(14.3)	(14.3)	(71.4)	(100.0)
5	Tea Garden	0	0	1	1
	Supervisory Staff	(0.00)	(0.00)	(100.0)	(100.0)
8	Non-Farm Wage	53	3	13	69
	Labour	(76.8)	(4.3)	(18.8)	(100.0)
10	Trade and Business	5	1	28	34
		(14.7)	(2.9)	(82.4)	(100.0)
11	Government Service	6	0	9	15
		(40.0)	(0.00)	(60.0)	(100.0)
12	Private Salaries Job	4	0	0	4
		(100.0)	(0.00)	(0.00)	(100.0)
14	Not Employed	2	1	6	9
		(22.2)	(11.1)	(66.7)	(100.0)
15	Others	17	0	3	20
		(85.0)	(0.00)	(15.0)	(100.0)
	Total	155	13	206	374
		(41.4)	(3.5)	(55.1)	(100.0)

Note: Figures in the parentheses refer to percentage to row totals.

changing the employer (Table 5.5). Outside the garden, a relatively larger proportion of workers would like to change their employers, but many among those who want to change their employers are casual workers of one type or other (Table 5.6). Given a chance nearly 5.6 per cent of workers in tea gardens and 12.3 per cent of those working outside the gardens would like to change their occupation (Tables 5.7 and 5.8). As already discussed, a relatively higher proportion of casual labourers in the tea gardens would like to change their occupation. This desire to change occupations is more widespread outside the garden than inside. It is the least among those working as permanent workers and supervisory staff in the tea gardens.

TABLE 5.5
Desire to Change Employer/Garden: Tea Garden

		Desire to Change Employer/Garden			
Code No.	Occupations	No	Yes	Not Applicable/ Can't Say	Total
1	Cultivators	0	1	12	13
		(0.00)	(7.7)	(92.3)	(100.0)
2	Agricultural Labour	3	0	0	3
		(100.0)	(0.00)	(0.00)	(100.0)
3	Tea garden Casual	21	11	2	34
	Labourer	(61.8)	(32.4)	(5.9)	(100.0)
4	Tea Garden	275	4	18	297
	Permanent Labourer	(92.6)	(1.3)	(6.1)	(100.0)
5	Tea Garden	6	1	0	7
	Supervisory Staff	(85.7)	(14.3)	(0.00)	(100.0)
8	Non-farm Casual	6	0	1	7
	Labour	(85.7)	(0.00)	(14.3)	(100.0)
10	Trade and Business	2	0	0	2
		(100.0)	(0.00)	(0.00)	(100.0)
14	Not Employed	n.a.	n.a.	n.a.	5
					(100.0)
15	Others	6	1	0	7
		(85.7)	(14.3)	(0.00)	(100.0)
	Total	335	19	21	375
		(89.3)	(5.1)	(5.6)	(100.0)

Note: Figures in the parentheses refer to percentage to row totals.

A related finding is that a significant proportion of tea garden workers do not think themselves as qualified to take up any other occupation. While this perception may not be entirely correct, the permanent tea garden labourers are the ones who mostly feel that they are not qualified for an occupational shift (Table 5.9). Very few of the tea gardens have actually tried to change their occupation (Tables 5.10 and 5.11). Among those who have tried to do so, casual tea garden workers come out as a prominent category. Casual workers in the non-farm sector have also tried to change their occupation to some extent.

Thus the overall findings from the perception analysis, notwithstanding its limitations point out that tea garden permanent worker have less inclination to change their occupations. On the other hand, those working as casual workers in tea gardens have

TABLE 5.6
Desire to Change Employer: Outside Tea Garden

		Would Like to Change the Present Employer			
Code No.	*Occupations*	*No*	*Yes*	*Not Applicable/ Can't Say*	*Total*
1	Cultivators	59	7	49	115
		(51.3)	(6.1)	(42.6)	(100.0)
2	Agricultural Labourers	16	12	57	85
		(18.8)	(14.1)	(67.1)	(100.0)
3	Tea garden Casual Labourer	8	3	4	15
		(53.3)	(20.0)	(26.7)	(100.0)
4	Tea Garden Permanent Labourer	5	1	1	7
		(71.4)	(14.3)	(14.3)	(100.0)
5	Tea Garden Supervisory Staff	1	0	0	1
		(100.0)	(0.00)	(0.00)	(100.0)
8	Non-Farm Wage Labour	49	14	6	69
		(71.0)	(20.3)	(8.7)	(100.0)
10	Trade and Business	15	2	17	34
		(44.1)	(5.9)	(50.0)	(100.0)
11	Government Service	9	0	6	15
		(60.0)	(0.00)	(40.0)	(100.0)
12	Private Salaries Job	3	1	0	4
		(75.0)	(25.0)	(0.00)	(100.0)
14	Not Employed	5	0	4	9
		(55.6)	(0.00)	(44.4)	(100.0)
15	Others	15	2	3	20
		(75.0)	(10.0)	(15.0)	(100.0)
	Total	185	42	147	374
		(49.5)	(11.2)	(39.3)	(100.0)

Note: Figures in the parentheses refer to percentage to row totals.

expressed their willingness to shift their occupation. A multiplicity of factors might have played a role in shaping these perceptions. In the following section an attempt is made to identify some of these underlying factors.

Factors Facilitating and Constraining Occupational Diversification

The process of livelihood diversification has attracted considerable analytical attention in recent literature. Households complicate and diversify the sources of their livelihood as a response to both threats

TABLE 5.7
Desire to Change Occupation: Tea Gardens

| Code No. | Occupations | *Given a Chance Would Like to Change Occupation* | | | |
		No	*Yes*	*Can't Say*	*Total*
1	Cultivators	9	0	4	13
		(69.2)	(0.00)	(30.8)	(100.0)
2	Agricultural	3	0	0	3
	Labourers	(100.0)	(0.00)	(0.00)	(100.0)
3	Tea garden Casual	14	11	9	34
	Labourer	(41.2)	(32.4)	(26.5)	(100.0)
4	Tea Garden	252	5	40	297
	Permanent Labourer	(84.8)	(1.7)	(13.5)	(100.0)
5	Tea Garden	5	1	1	7
	Supervisory Staff	(71.4)	(14.3)	(14.3)	(100.0)
8	Non-Farm	2	1	4	7
	Wage Labour	(28.6)	(14.3)	(57.1)	(100.0)
10	Trade and Business	1	1	0	2
		(50.0)	(50.0)	(0.00)	(100.0)
14	Not Employed	n.a.	n.a.	n.a.	5
					(100.0)
15	Others	1	1	5	7
		(14.3)	(14.3)	(71.4)	(100.0)
	Total	291	21	63	375
		(77.6)	(5.6)	(16.8)	(100.0)

Note: Figures in the parentheses refer to percentage to row totals.

and opportunities. Livelihood comprises the capabilities, assets (including both material and social resources) and activities required for a means of living. A livelihood 'encompasses income, both in cash and kind, as well as social institutions (kin, family, compound, village and so on), gender relations and property rights to support and to sustain a given standard of living' (Ellis 1998). Thus, conceptually the livelihood approach goes beyond the usual emphasis on income and employment and brings in social institutions, which play a crucial role in determining the constraints and options of individuals and households.[2] *Livelihood diversification* is defined as 'the process by

---

[2] Social institutions, which include rules of conduct, norms and expected behavioural outcomes, critically condition the access of households to land, common property resources (CPRs) and other tangible and non-tangible resources. Social norms on permissible courses of action for women, for example, may influence the livelihood options available for women compared to men (Ellis 1998).

TABLE 5.8
Desire to Change Occupation: Outside Tea Garden

Code No.	Occupations	*Desire to Change Occupation*			
		No	*Yes*	*Not Applicable/ Can't Say*	*Total*
1	Cultivators	58	8	49	115
		(50.4)	(7.0)	(42.6)	(100.0)
2	Agricultural	14	13	58	85
	Labourers	(16.5)	(15.3)	(68.2)	(100.0)
3	Tea garden Casual	8	3	4	15
	Labourer	(53.3)	(20.0)	(26.7)	(100.0)
4	Tea Garden	5	1	1	7
	Permanent Labourer	(71.4)	(14.3)	(14.3)	(100.0)
5	Tea Garden	1	0	0	1
	Supervisory Staff	(100.0)	(0.00)	(0.00)	(100.0)
8	Non-Farm Wage	49	14	6	69
	Labour	(71.0)	(20.3)	(8.7)	(100.0)
10	Trade and Business	14	3	17	34
		(41.2)	(8.8)	(50.0)	(100.0)
11	Government Service	8	1	6	15
		(53.3)	(6.7)	(40.0)	(100.0)
12	Private Salaries Job	3	1	0	4
		(75.0)	(25.0)	(0.00)	(100.0)
15	Others	15	2	3	20
		(75.0)	(10.0)	(15.0)	(100.0)
	Total	180	46	148	374
		(48.1)	(12.3)	(39.6)	(100.0)

Note: Figures in the parentheses refer to percentage to row totals.

which households construct a diverse portfolio of activities and social support capabilities for survival and in order to improve their standard of living' (ibid.). 'Considerations of risk spreading, consumption smoothing, labour allocation smoothing, credit market failures, and coping with shocks can contribute to the adoption, and adaptation over time, of diverse rural livelihoods' (Ellis 1999). Livelihood diversification strategies are followed both by the poor as well as the non-poor. To put it another way, livelihood diversification could be a part of the survival or accumulative strategies of households. Occasionally, a diversification strategy opted for during a period of crisis could form part of an accumulative strategy. Analytically, livelihood diversification could be distinguished both in terms of its nature and implications. Faced with adverse livelihood prospects,

TABLE 5.9
Perception Regarding Qualification to Change Occupation: Tea Gardens

Code No.	Occupations	Do you think to be qualified to work in other occupation			
		No	*Yes*	*Can't Say*	*Total*
1	Cultivators	8	0	5	13
		(61.5)	(0.00)	(38.5)	(100.0)
2	Agricultural	3	0	0	3
	Labourers	(100.0)	(0.00)	(0.00)	(100.0)
3	Tea garden Casual	17	6	11	34
	Labourer	(50.0)	(17.6)	(32.4)	(100.0)
4	Tea Garden	250	3	44	297
	Permanent Labourer	(84.2)	(1.0)	(14.8)	(100.0)
5	Tea Garden	5	1	1	7
	Supervisory Staff	(71.4)	(14.3)	(14.3)	(100.0)
8	Non-Farm Wage	3	0	4	7
	Labour	(42.9)	(0.00)	(57.1)	(100.0)
10	Trade and Business	2	0	0	2
		(100.0)	(0.00)	(0.00)	(100.0)
14	Not Employed	3	2	0	5
		(60.0)	(40.0)	(0.00)	(100.0)
15	Others	1	0	6	7
		(14.3)	(0.00)	(85.7)	(100.0)
	Total	292	12	71	375
		(77.9)	(3.2)	(18.9)	(100.0)

Note: Figures in the parentheses refer to percentage to row totals.

often households diversify the sources of their livelihood under distress. The survival motive and push factors play a crucial role in such situations. Such livelihood diversification is often induced by loss of vital livelihood assets or endowments of the households. At times, it becomes a response to catastrophic changes in the economic or ecological basis of survival of the households. It has also been suggested that in high risk economic and ecological contexts households tend to diversify their sources of livelihood in an attempt to minimise risks and also to take advantage of slack resources of various kinds. However, livelihood diversification forms part of the *strategies of accumulation* of the households as well. Households tend to diversify their sources of livelihood to increase their earnings, in response to existing and emerging opportunities. The factors which triggers such diversification include higher return on labour, higher

TABLE 5.10
Attempts to Change Occupation: Tea Gardens

Code No.	Occupations	Ever tried to change occupation			
		No	Yes	Can't Say	Total
1	Cultivators	7	0	6	13
		(53.8)	(0.00)	(46.2)	(100.0)
2	Agricultural Labourers	2	0	1	3
		(66.7)	(0.00)	(33.3)	(100.0)
3	Tea garden Casual Labourer	13	5	16	34
		(38.2)	(14.7)	(47.1)	(100.0)
4	Tea Garden Permanent Labourer	196	6	95	297
		(66.0)	(2.0)	(32.0)	(100.0)
5	Tea Garden Supervisory Staff	2	1	4	7
		(28.6)	(14.3)	(57.1)	(100.0)
8	Non-Farm Wage Labour	4	0	3	7
		(57.1)	(0.00)	(42.9)	(100.0)
10	Trade and Business	1	1	0	2
		(50.0)	(50.0)	(0.00)	(100.0)
15	Others	1	0	6	7
		(14.3)	(0.00)	(85.7)	(100.0)
	Total	229	15	131	375
		(61.1)	(4.0)	(34.9)	(100.0)

Note: Figures in the parentheses refer to percentage to row totals.

return on investments, lower risk of some activities compared to traditional occupations, generation of cash in order to meet household objectives, economic opportunities, often associated with social advantages, offered in urban centres and outside of the region or country, appeal of urban life, in particular to younger people.

Livelihood diversification is generally conceptualised as a broader process than simply income and employment diversification. Here our focus is on employment rather than on livelihood. Nevertheless, the literature on livelihood diversification points out some significant constraints that limit or affect the capabilities of households and individuals to diversify their sources of earning, even in the context of expanding opportunities in certain segments of the labour markets. The capabilities of households and individuals to take advantage of better earning opportunities, for example, might be limited by their lack of access to certain crucial assets (natural, physical, financial, human social capitals) or by formal or informal entry barriers operating at various levels. The livelihood research has

TABLE 5.11
Attempts to Change Occupation: Outside Tea Garden

Code No.	Occupations	No	Yes	Can't Say	Total
		Attempted to Change Occupation			
1	Cultivators	59	7	49	115
		(51.3)	(6.1)	(42.6)	(100.0)
2	Agricultural Labourers	19	8	58	85
		(22.4)	(9.4)	(68.2)	(100.0)
3	Tea garden Casual Labourer	8	3	4	15
		(53.3)	(20.0)	(26.7)	(100.0)
4	Tea Garden Permanent Labourer	5	1	1	7
		(71.4)	(14.3)	(14.3)	(100.0)
5	Tea Garden Supervisory Staff	1	0	0	1
		(100.0)	(0.00)	(0.00)	(100.0)
8	Non-Farm Wage Labour	50	13	6	69
		(72.5)	(18.8)	(8.7)	(100.0)
10	Trade and Business	15	3	16	34
		(44.1)	(8.8)	(47.1)	(100.0)
11	Government Service	9	0	6	15
		(60.0)	(0.00)	(40.0)	(100.0)
12	Private Salaries Job	3	1	0	4
		(75.0)	(25.0)	(0.00)	(100.0)
15	Others	15	2	3	20
		(75.0)	(10.0)	(15.0)	(100.0)
	Total	188	39	147	374
		(50.3)	(10.4)	(39.3)	(100.0)

Note: Figures in the parentheses refer to percentage to row totals.

been criticised on the grounds that it does not pay adequate attention to the political economy of unequal opportunities to acquire assets, skills and institutions (Murray 2001; De Haan and Zoomers 2005). The resources that individuals (and households) can access depends not only on the local economy and ecology but also on the political economy of resource sharing and distribution among various groups of claimants. As De Haan and Zoomers (2005: 34) point out:

> Failed access and the resultant poverty or social exclusion can also be the result of a mechanism by which some people exclude others from access to resources, with the objective of maximising their own returns…They use property relations or certain social or physical characteristics such as race, gender, language, ethnicity, origin or religion to legitimize this fencing-in of opportunities. Social exclusion and poverty are then consequences of social closure, a form

of collective action which gives rise to social categories of eligibles and ineligibles...Livelihood activities are not neutral, but engender process of inclusion and exclusion.

Further, the access to resources of households does not mean automatic and equitable access for all household members. Women, children, disabled and aged persons often do not have the same access as males. Thus, intra household distribution of power, privilege and authority determines the disparity in access to these 'available' resources as well as their effective utilisation by the members (Mishra 2007).

In many ways the 'tea garden labour' community has a unique history and their occupational diversification is driven by considerations unique to their specific local contexts. As our previous investigations point out, there has indeed been very little occupational diversification among the households living within the gardens. Even non-significant portions of the workers who belong to ex-tea garden labour households continue to find employment within the gardens. Further, our field survey analysis points out that occupational diversification and mobility is relatively higher among the current generation of workers than their fathers and grandfathers. What needs to be emphasised here is that a substantial portion of those who have moved out of the tea gardens have gone to agriculture, either as cultivators or as agricultural labourers. Thus even when there is a degree of diversification present, it is mostly within the agricultural sector itself. Diversification to non-farm occupations was not very significant and the access to salaried permanent jobs in the service sector was negligible. Research evidence from a number of countries suggests that access to rural non-farm employment and income is largely determined by factors like education, social capital, ethnicity and caste, gender dynamics, financial capital, and physical infrastructure and information. *Education* is crucial because many of the better paid jobs require formal schooling (Ferreira and Lanjouw 2001). While education may not be an important factor in determining access to casual non-agricultural wage employment, for regular, salaried employment it is clear that probability of employment rises as education levels rise (Lanjouw 1999). Examining the returns to education within the rural non-farm sector, Jolliffee (1999, cited in Lanjouw 1999) reports that earnings tend to rise sharply with higher education levels. However, the extent to which schooling, beyond basic literacy and primary level, provides skills that matter in the majority of non-farm activities is still doubtful (Davis 2003). *Social capital*, defined as the degree of interaction with others in the context of social networks (Fafchamps and Minten 1999), can act as a catalytic factor influencing access to non-farm occupations. It can enable economic agents to reduce transaction costs and partially circumvent entry barriers arising from imperfect markets. Among the many advantages of social capital, are 'the access to relevant market information and buyers, wage employment and business opportunities, formal and informal loans, cash advances, inputs on credit, skills, shared resources for production and marketing, and migration opportunities'(Davis 2003: 11). Studies on India's informal economies have often stressed the role of social networks and shared social identity in defining access to employment and information (Harriss-White 2003). Moreover, the concentration of critical social capital in the hands of a privileged minority may act as a barrier to entry to employment in lucrative sectors for a majority of the rural population. *Ethnicity* is an important determinant of access to better paid jobs outside agriculture, and can play both a facilitating and constraining role. Higher caste status, given the unequal distribution of power, privilege and assets in the local economy, often opens up the scope for better earning opportunities, through better access to education and information. The role of informal networks of mutual support and insurance, in some contexts, may act as a facilitator for participation in specific non-farm occupations and activities.[3]

Inadequate access to capital is one of the most widely cited obstacles to growth of entrepreneurship and investment (Davis 2003). However, demand constraints might also be operative in restricted access to *credit*. Since the access to formal credit remains highly restricted, in rural areas, those with some access to credit, formal or informal, have a greater probability of entering the non-farm sector, particularly as self-employed persons. Physical

---

[3] Upadhyay and Mishra (2004), in a field survey on micro-enterprises in Arunachal Pradesh, for example, note that ethnic networks have played a significant role in labour recruitment in skill-based occupations and also in sourcing raw materials and finished products in trading enterprises.

infrastructure, particularly road connectivity, enhances the scope for participation in non-farm activities. Scope for skill formation and supply side constraints in input markets are linked to the availability of infrastructure. The influence of these factors works conjointly rather than individually. Assets have a degree of interdependence and fungibility (Ellis 2000).

Occupational diversification of workers in the tea plantation depends on two sets of variables that affect the decision to work in specific sectors: (*a*) the willingness or desire to change the occupation and (*b*) the capacity or ability to change the occupation. First, there could be a number of reasons for which the tea labour households, given a chance, would like to continue to work within the garden itself. For example, decisions to change occupations are often taken at the household level rather than at the individual level. It is possible that in spite of the wage rate within the garden being lower than the expected wage rate outside the garden for a specific individual, the total *family* earnings outside the garden might be lower than what the household would earn inside the garden, mainly because a higher number of members could get employed within the garden. In such cases there would be no incentive for labour to move out of the plantation sector even if the wage rate was higher outside. Therefore, it is the total household income and not the wage rate as such that influences the labourers' decision.

There could be various constraints that limit the employment choices of the tea garden workers as well. For long the 'unfreedom' and bondage in labour regimes that had been a part of the early history of the development of the plantation sector has had an impact on the labour relations in the plantations.[4] The pre-capitalist/feudal base in the structure of relation between labour and capital/management in the gardens (Behal and Mohapatra 1992) might still be influencing labour relations to some extent. This might also have acted as an impediment to the emergence of labour as an independent factor of production, as in a capitalist mode of production.

---

[4] The 'industrial plantation' system that emerged in the nineteenth century represented 'new forms of capital', but its link with unfreedom of labour remains one of the problematic issues in understanding the nature of labour processes in plantations (Brass and Bernstein 1992: 4–5).

162 of 264 ■ Unfolding Crisis in Assam's Tea Plantations

Access to Education

One of the key constraints faced by those who want to move into other occupations is lack of education and skills. Education has long been associated with upward social mobility. But the tea garden labourers and their children face a number of hurdles in getting access to better education. While there is the provision for schools in most of the gardens, more often than not these schools offer education only up to the primary or upper primary levels. For anything higher than this children have to travel a long distance. The problems are acute as most of the gardens are located in isolated pockets with poor connectivity to nearby urban centres and villages. Even when schools are available there is virtually no public transport available for the children. Although the gardens have primary schools, these schools offer little in terms of the nature and quality of schooling; these schools offer hardly any substantial learning to the pupils. Our survey team came across single-teacher primary schools, where parents felt that there was hardly any teaching going on for months. In the cases where teachers were provided by the tea garden management, the teachers alleged that they were overburdened with work other than that involving running the perpetually understaffed schools. On the demand side, the single most important factor contributing to the lack of interest in schooling is the availability of jobs within gardens as 'non-adult' workers. It suits both the parents and the management, but limits the scope of learning and occupational diversification among the workers. As children start supplementing the family income at an early stage it lessens the burden of the parents. In our group discussions many parents in fact demanded that their children be allowed to work in the gardens. Their argument was that since there were hardly any jobs in the tea gardens it was better for their children to get trained in their 'ancestral' occupation. The mere availability of such an option to earn at an early age however makes formal education less attractive even for the children (Fernades et al. 2003). From the management's point of view such an arrangement creates a captive labour force whose opportunity cost of being employed is nearly zero.[5] The all

---

[5] Although evidence at the grass roots suggests that recruitment of all categories of labourers has suffered in recent years and so has the recruitment of child labour.

too familiar 'tea leaves plucking as a way of life argument' was also cited by a number of persons whom we interviewed. At least in one case, a senior level manager with long experience in the sector raised serious doubts regarding the capability of the tribal labour community to be educated and absorbed anywhere else. Of course, during the course of our discussion he mentioned that the 'half-educated lot' among the labour community were the real troublemakers and it is they who were responsible for breaking the harmony of the tea gardens, whatever the later means.

It is however, clear from Table 4.23 in Chapter 4 that an educational attainment among the tea garden workers is abysmally low. On other hand, there is a significant gap between the mean years of schooling between the labourers and those who have joined white-collar professions, inside or outside the tea gardens. The average years of schooling among those who work within the gardens and those who work outside, however, is not significantly different (Table 5.12). The reason for this is that the non-tea garden occupations are also in the low-income end of the job spectrum. There is hardly

TABLE 5.12
Occupation-wise Mean Years of Schooling

		Mean Years of Schooling	
Sl. No	*Occupations*	*Inside Gardens*	*Outside Gardens*
1	Tea Garden Casual Labour	2	2
2	Tea Garden Permanent Labour	2	3
3	Tea Garden Lower Supervisory Staff & Service Providers	6	5
4	Tea Garden Clerical Staff	13	10
5	Cultivators	4	4
6	Agricultural Labourers	4	2
7	Non-farm Casual Labour (unskilled)	3	4
8	Non-farm Casual Labour (skilled)	6	4
9	Trade and Business	8	7
10	Government Service	9	11
11	Private Salaried Service	9	8
	All Tea	2	3
	All Non Tea	4	4
	Total	3	4

any skill formation or access to vocational education for the workers, so the best they can hope for is options in the casual labour market, particularly outside agriculture. To some extent ownership of land, mostly purchased out of the retirement benefits, influences the choice of cultivation as an occupation.

The Nature of the Local Economy

Another significant factor influencing the occupational diversification process in the vicinity of the tea gardens is the nature of the local economy. Our field research in the three gardens chosen for intensive study and elsewhere brings out the significance of the spatial context of the economy as a key determinant of the extent and nature of labour movement from the tea gardens to the other sectors of the local economy. Tea gardens not only differ in their sizes and the volume of tea leaves produced but also in terms of their locations. Relatively larger tea gardens, located along the national highways are, more often than not are more closely integrated with the local economy. The agricultural goods produced within and outside the gardens flows into the local markets more easily and also the purchasing power infused through wages and salaries in the gardens sustains, to some extent, the local non-farm economy, particularly retail trading of various kinds. These linkages are relatively weak in remote gardens with poor connectivity status. These gardens operate with varying degrees of isolation from the local economy. The flow of labour to and from the tea gardens is as such less robust than that of retail goods (like daily consumables) and local products (particularly grain, dairy products, etc.). In relatively isolated tea gardens the options for labour remain severely limited as agriculture itself remains low productive and is carried out mostly for subsistence. There is hardly any institutional support for adoption of better technology and distance from markets makes commercial agriculture non-viable. The policy of the tea gardens regarding allotment of land remains one of the factors that has historically facilitated or constrained the scope for the labourers to have land for cultivations. While in recent times most tea gardens do not have a policy of providing surplus land to their labour force, in the past there was a liberal attitude towards this. So many labourer families got access to cultivate unutilised land within or outside the tea garden. The legal status of all this land may not always be clear but cultivation of this kind did provide a sort of livelihood security to the labourers. A lot of the land owned by the labourers was reported as land acquired through clearing of forests. Many informants indicated that there has been a gradual alienation of land from the tea tribes to others in recent years through the classic process of debt-induced alienation of land. Another source of land has been purchase through the land market. It is not uncommon for the labourers to utilise their savings to purchase land in the nearby villages and shift to cultivation after retirement.

In Lakhimpur district we found that agriculture and the agricultural labour market remain the main channels of occupational diversification. Households also reported semi-permanent migration to the non-farm and informal sector jobs in neighbouring Arunachal Pradesh as another source of livelihood in difficult times. In tea gardens located near Dibrugarh, however, jobs in the urban labour markets provided the best opportunities. Many have joined these jobs as rickshaw pullers, domestic servants and construction labourers. Those who have some skills through learning from peers and elders within the tea gardens have joined as skilled workers in construction, electricity, etc. A few have taken the more formal route of joining vocational institutes. But, over all, the easy access to the urban labour market has facilitated the diversification process to a great deal. In the relatively isolated tea gardens of Sivasagar district, the scope for such diversification and mobility was found to be severely limited.

The Ethnic Dimension

The political scenario in Assam, as in many other northeastern states, has been deeply influenced by the politics of identity and difference (Vandekerckhove 2009). The question of migration to Assam has for long been one of the emotive factors in the politics of Assam. Thus the perceived 'otherness' of the tea garden labour force is also a factor influencing the occupational diversification strategies of the households. The integration of the large tea labour community with the local population has been very slow.[6]

---

[6] This is not to argue that there are no other kinds of social differentiation within the gardens. Beteille (1981: vii) brings out the diversity within the garden labour force and its implications succinctly. 'The social system of the

One of the key factors influencing this relatively slower pace of integration is perhaps the (relatively) self-sufficient economy in the gardens. The labour market in the tea garden areas remains deeply fragmented in this aspect. The labour outside the gardens, unless they belong to the 'the garden community', hardly compete with the garden labourers for the employment available in the gardens. Similarly, the flow of labour from the gardens to the local labour markets, though increasing over a period of time, remains constrained. The two important channels through which this flow continues are: (*a*) the assimilation of a section of garden labourers, or the members of what is called the 'tea labour community' into the nearby villages as cultivators, which has further opened up the scope for work as agricultural labourers; and (*b*) the access to jobs in the urban informal sectors. But one of the factors, which have influenced the scope for diversification, is the access to homestead land. Households, which own some land, have found it easier to opt for jobs outside the gardens, but those who do not have such land generally prefer to stay inside the gardens and try to supplement their earnings from the garden with outside income. As it is becoming increasingly difficult to get permanent jobs in the gardens, many households have tried to diversify their sources of earnings, but without shifting their base from the gardens. This, at least partially explains the presence of workers with non-tea occupations within the gardens. When the households opt for moving out they prefer to move out to localities or villages with a tea labour community. One of the characteristics of migrant labour in the plantation sector is: historically they were brought in large numbers from one locality. This might have resulted in strong social bonding and kinship and

---

plantation is not exhausted by the organisation of work within it. For, while the labourers are homogeneous from the viewpoint of work and wages, they are heterogeneous in other very important regards'. Referring to the differences among Nepali and tribal workers in the Dooars plantations, he argues that 'although migration to the plantations disrupted many traditional bonds, the division between tribes continues to be a feature of life on the plantations. The disjunction between the "homogeneous economic activity" of the workers and "heterogeneous social life" provides the backdrop, as it were, to the problem — theoretical as well as practical — of class formation on the plantations'.

hence might have acted as an impediment to their free decision to move out for job.

Social Security

Behind the reluctance of individuals and households to opt for an occupation outside the garden lies the significance attached to the garden as a 'way of life' and as a 'home'. For example, one of the workers pointed out to the researchers during an interview: 'It's our world. We do not know any other world'. Members of the managerial staff also expressed similar views. However behind these 'tea garden-as-way-of-life' arguments there could be economic reasons for not moving out of the tea garden.

Although there were several complaints about payments being delayed or denied, by and large, the provisions of the Tea Labour Act provides some amount of social security to the workers. The provision of ration as part of wages acts as an added incentive for the households. Casual workers did complain about the denial of rations through denial of continuous days of work. Many recently retired workers complained of harassment and denial of their rights. But wherever strong and active trade union leaders were present, it was noticed, workers received their dues. Thus, in spite of some leakages and non-observation of provisions of the act at the ground level, permanent workers in the garden do have access to some social security provisions that are better than what these workers might have in alternative occupations available to them.

An attempt has been made to calculate a comparable social security score to all categories of workers. Mean Social Security score has been calculated in the following way. The following aspects of social security have been taken into account and a final score was prepared: S_1=Paid leave; S_2=Provident fund; S_3=Pension; S_4=Medical support; S_5=Housing support; S_6=Maternity leave; S_7=Insurance; S_8=Injury allowance; S_9=Other benefit. Access to any of these social security benefits is given the value 1, and its non-availability is assigned a value of zero. The Composite Social Security Score (S) $= \sum S_{i, i = 1...9}$.

It is found that supervisory staff and permanent workers in the gardens have an overall higher score in social security than all the other categories of workers (Table 5.13). As expected, cultivators,

TABLE 5.13
Occupation-wise Mean Social Security Score

Code No.	Occupations	Mean Social Security Score	
		Inside tea Garden	Outside Tea Garden
1	Cultivators	0.54	0.04
2	Agricultural Labourers	0.00	0.00
3	Tea garden Casual Labourer	2.79	0.40
4	Tea Garden Permanent Labourer	4.14	3.57
5	Tea Garden Supervisory Staff	4.89	5.00
8	Non-Farm Wage Labour	0.43	0.17
10	Trade and Business	2.50	0.05
11	Government Service	–	2.67
12	Private Salaries Job	–	1.00
15	Others	4.29	0.20

agricultural labourers and non-farm wage labourers have the least social security score. Tea garden casual labourers do not have the same access to social security as the permanent workers, but still, their social security score is higher than that of casual workers in other sectors. This, in the final analysis, makes the tea garden job more lucrative than other alternatives for the workers.

Conditions of Living

The condition under which labour is being reproduced is an important aspect of the production conditions in the gardens. As argued earlier, while the demand for labour in the gardens has been responding to the forces of the market, it is the supply of labour that gets determined through the specificities of the tea sector. The historical continuity of the enclave economy creates specific features of labour supply. There are several aspects of the plantation economy that defines the linkages between living and working conditions.

> In the plantations, to a greater extent than a factory, the management exercises direct control over the labourers not only at work but also outside work. This is because the labourers have no choice but to live on the plantation in accommodation provided by the employer. This created the condition for the paternalism characteristic of the

plantation system. But we must understand this paternalism for what it was: it was paternalism between people who were divided not only in their interest but also in their ideas and values. As such, it was far more oppressive in nature than the paternalism characteristic of the traditional village community (Beteille 1981: vii–viii).

The relative immobility of plantation labour, which has been clearly brought out in the earlier chapters, is at least partly rooted in the arrangements for reproduction of labour within this enclave economy. The linkages between conditions of living and working have been well established in labour studies, but the production organisation of tea labour gardens brings out these linkages in bold relief.

The study of living conditions inside and outside the gardens that has been attempted here is of significance for two reasons. In a descriptive sense, it brings out the conditions of labour working in the tea gardens. This is of added importance, because, in a nutshell, it captures the (in)effectiveness of elaborate legislation on the rights of the workers. At another level, it exposes the limitations of labour standards as a tool to improve the conditions of workers through visions of enlightened consumer activism in a globalising world. But, more importantly perhaps, in an analytical sense, the study of living conditions informs the underlying mechanisms through which labour relations becomes 'flexible' and informalised within the overall edifice of an organised industry. In its response to global competitions, falling profits and rising costs, the tea industry has deliberately attempted to subvert legislation on workers' rights by casualising a substantial section of the workforce, but this has not resulted in a spatial or sectoral reallocation of the labour force. Rather a 'reserve army of casual workers' continues to stay and survive within and alongside the gardens and this is crucial in maintaining the flexibility in labour employment without increasing the transaction and production costs of employing more labour during periods of boom and expansion. Work is not only hierarchical but also gendered (Bhowmik 1981; Reddock and Jain 1998; Jain 1998; Nagraj and Vedavalli 2004). Thus, the linkages between the living and working conditions of labour in the gardens are structural rather than sporadic. Further, such linkages are embedded within the multiple subjectivities based on gender, kinship and language that permeate the lifeworlds and worlds of work of labour in the tea gardens.

We start by looking at some of the basic demographic features of the labour households, and then move on to the analysis of access to basic amenities within and outside the gardens. The analysis of household consumption and entitlements to basic items such as food and shelter is expected to bring out the precariousness of the livelihood base of the labour households.

Demographic Characteristics of Labour Households

The basic demographic features of the sample households show a relatively large family size, broadly comparable to rural India (Table 5.14). However, the interrelationship between working conditions and household dynamics could be inferred from the facts that, by and large, the households inside the gardens have a higher sex ratio, a higher dependency burden. On the other hand, among the labour households residing outside the gardens, the share of working adults as well as the share of female-headed households is higher. As discussed earlier, the process of moving out of the gardens has got several interfaces with the work and employment status of the individuals. Households or individuals typically move out when they cannot secure employment within the gardens any longer and also cannot reside as part of the extended family. Typically, young couples, when they fail to get employment within the gardens or old couples after their retirement, choose to move out of the gardens. During the field survey we came across numerous instances where the shrinking labour market in the tea sector had a bearing upon interpersonal relations within the families. As no new recruitments for permanent workers are taking place, and only one son or daughter of a old employee gets a chance to substitute the parents, competition, jealousy and rivalry among siblings is reported to have intensified. Intergenerational relationships have suffered as well. Working parents approaching their retirements are under intense pressure to relinquish their jobs in favour of one of their children and this has often resulted in increased intra-family conflicts and greater vulnerability of older parents after retirement.

TABLE 5.14

Demographic Characteristics of Labour Households

Sl. No.	Characteristics	Districts						All	
		1		2		3			
		Within Gardens	Outside Gardens	Within Gardens	Outside Gardens	Within Gardens	Outside Gardens	Within Gardens	Outside Gardens
1	Average Household-size	4.34	4.47	5.63	5.40	5.51	4.93	5.16	4.93
2	Sex Ratio	950	911	1012	971	1038	984	1003	957
3	Child Dependency Ratio	61.11	64.37	70.26	64.36	74.05	65.30	68.56	64.70
4	Dependency Ratio	63.45	64.97	71.28	65.56	79.30	66.75	71.35	65.81
5	No. of Working Age Adults per Households	2.73	2.67	3.15	3.40	2.72	3.32	2.87	3.10
6	Share of Female-Headed Households (in percentages)	8.00	19.20	6.45	5.65	10.32	19.20	8.27	14.71

Access to Basic Amenities

The ownership of dwellings and their conditions, which are normally taken as indicators of household well-being, have additional significance for tea garden labourers. As spatial location of the settlement is linked to community membership and associated tangible and non-tangible benefits, particularly among immigrant communities working in an enclave economy, tea garden labourers, given a choice, do not like to move out of the gardens. However, as a result of increase in family sizes and inadequate housing provisioning within the gardens, 22 per cent of the labour households within the gardens do not have access to employer-provided housing facility (Table 5.15). Most of those who are not residing within the gardens have ownership over their dwelling units. This is partly because of the fact that when tea labour families move out of the gardens, they do not seek jobs or residences randomly. Typically, they prefer to move to villages already having some tea-labour/adivasi households. However, by and large, the quality of housing is poorer among those who have shifted to locations outside the gardens (Table 5.16). Rooms per houses are slightly higher outside, and there is higher congestion in dwellings inside the gardens (Table 5.17).

So far as access to drinking water is concerned, it was found that within the gardens only 10 per cent of households have access to tap water, while 78 per cent depend on the tube well (Table 5.18). Outside the gardens, the dependence on wells for drinking water was substantially higher. Thus, although more households within the gardens have to share the water source and a greater share of households residing outside the gardens have the water source within their premises, the drinking water condition for those living outside is worse than that of households within the gardens.

Within the gardens, 73 per cent of the households depend on kerosene for lighting purposes, and there was hardly any inter-district variations in this (Table 5.19). However, conditions outside the gardens were even worse. So far as domestic energy-use pattern is concerned, most of the labour households depended upon firewood; the dependence on firewood seems to be higher within the garden than outside it (Table 5.20). Most tea gardens allow labour households to collect fuel wood from within the gardens. Thus,

TABLE 5.15
Ownership of Dwellings

		Districts						All	
		1		2		3			
Sl. No	Ownership Type	Within Gardens	Outside Gardens	Within Gardens	Outside Gardens	Within Gardens	Outside Gardens	Within Gardens	Outside Gardens
1	Owned	33 (26.4)	122 (97.6)	9 (7.3)	116 (93.5)	17 (13.5)	125 (100)	59 (15.7)	363 (97.1)
2	Provided by Garden	83 (66.4)	0 (0.0)	100 (80.6)	0 (0.0)	108 (85.7)	0 (0.0)	291 (77.6)	0 (0.0)
3	Homestead by Garden	0 (0.0)	0 (0.0)	0 (0.0)	5 (4.0)	0 (0.0)	0 (0.0)	0 (0.0)	5 (1.3)
4	Rented	7 (5.6)	0 (0.0)	13 (10.5)	1 (0.08)	0 (0.0)	0 (0.0)	20 (5.3)	1 (0.3)
5	Other	2 (1.6)	3 (2.4)	2 (1.6)	2 (1.6)	1 (0.8)	0 (0.0)	5 (1.3)	5 (1.3)
	Total	125 (100)	125 (100)	124 (100)	124 (100)	126 (100)	125 (100)	375 (100)	374 (100)

TABLE 5.16
Types of Dwellings

Dwellings	Types	Within Gardens	Outside Gardens	All
Roof	Thatched	22	116	138
		(5.9)	(31.0)	(18.42)
	Tin shed	344	238	582
		(91.7)	(63.6)	(77.70)
	RCC	3	8	11
		(0.8)	(2.1)	(1.47)
	Others	6	12	18
		(1.7)	(3.2)	(2.40)
Wall	Mud	93	276	369
		(24.8)	(73.8)	(49.27)
	Assam type	7	16	23
		(1.9)	(4.3)	(3.07)
	Pucca	268	22	290
		(71.5)	(5.9)	(38.72)
	Other	7	60	67
		(1.87)	(16.0)	(8.92)
Floor	Kutcha	327	338	665
		(87.2)	(90.4)	(88.79)
	Semi-Pucca	11	9	20
		(2.9)	(2.4)	(2.67)
	Pucca	36	26	62
		(9.6)	(7.0)	(8.28)
	Not	1	1	2
	Reported	(0.3)	(0.3)	(0.27)

Table 5.17
Persons per Dwellings

Districts	Persons per Rooms			Rooms per Dwellings		
	Inside Gardens	Outside Gardens	Total	Inside Gardens	Outside Gardens	Total
1	0.72	1.27	1.32	3.12	3.51	3.32
2	1.72	1.40	1.55	3.27	3.85	3.56
3	1.92	1.82	1.87	2.86	2.70	2.78
All	1.67	1.60	1.57	3.08	3.35	3.22

Note: Kitchens and small, enclosed extensions on verandas have also been counted as 'rooms'.

Table 5.18
Access to Drinking Water

	Districts						All	
	1		2		3			
Characteristics	Within Gardens	Outside Gardens	Within Gardens	Outside Gardens	Within Gardens	Outside Gardens	Within Gardens	Outside Gardens
Source of Drinking Water								
Tap	13	11	0	1	24	0	37	12
	(10.40)	(8.80)	(0.00)	(0.81)	(19.05)	(0.00)	(9.9)	(3.2)
Well	5	25	1	17	41	100	47	142
	(4.00)	(20.0)	(0.81)	(13.71)	(32.54)	(80.0)	(12.5)	(38.0)
Tube Well	107	69	123	106	61	11	291	186
	(85.60)	(55.20)	(99.19)	(85.48)	(48.41)	(8.80)	(77.6)	(49.7)
Pond/Reservoir	0	20	0	0	0	3	0	23
	(0.00)	(16.00)	(0.00)	(0.00)	(0.00)	(2.40)	(0.00)	(6.1)
River/Canal	0	0	0	0	0	11	0	11
	(0.00)	(0.00)	(0.00)	(0.00)	(0.00)	(8.80)	(0.00)	(2.9)
Shared Sources								
Personal	18	56	65	97	5	20	88	173
	(14.4)	(44.8)	(52.4)	(78.2)	(4.0)	(16.0)	(23.5)	(46.3)
Shared	105	69	59	27	121	105	285	201
	(84.0)	(55.2)	(47.6)	(21.8)	(96.0)	(84.0)	(76.0)	(53.7)
Not Reported	2	0	0	0	0	0	2	0
	(1.6)	(0.00)	(0.0)	(0.00)	(0.0)	(0.00)	(0.5)	(0.00)

(Continued...)

(Table 5.18 Continued..)

	Districts							
	1		2		3		All	
Characteristics	Within Gardens	Outside Gardens	Within Gardens	Outside Gardens	Within Gardens	Outside Gardens	Within Gardens	Outside Gardens
Distance from Source								
Within the dwelling	0	43	9	84	0	42	9	169
	(0.00)	(34.4)	(7.3)	(67.74)	(0.00)	(33.60)	(2.4)	(45.19)
Less than 100 meters	6	40	47	16	13	50	66	106
	(4.8)	(32.0)	(37.9)	(12.90)	(11.3)	(40.00)	(17.6)	(28.34)
> 100 m but less than 500 meters	31	16	37	2	98	25	166	43
	(24.8)	(12.8)	(29.8)	(1.61)	(77.8)	(20.00)	(44.3)	(11.5)
>500 m but less than 1 km	81	10	31	11	14	5	126	26
	(64.8)	(8.00)	(25.0)	(8.87)	(11.1)	(4.00)	(33.6)	(6.95)
>1 km but more than 1 km	5	16	0	11	1	3	6	30
	(4.0)	(12.80)	(0.00)	(8.87)	(0.8)	(2.40)	(1.6)	(8.02)
Not Reported	2	0	0	0	0	0	2	0
	(1.6)	(0.00)	(0.00)	(0.00)	(0.00)	(0.00)	(0.53)	(0.00)

TABLE 5.19
Source of Lighting

Source	Within the Gardens				Outside the Gardens				All
	District			Total				Total	
	1	2	3		1	2	3		
No lighting	3	2	0	5	1	1	1	3	8
	(2.40)	(1.61)	(0.00)	(1.33)	(0.80)	(0.81)	(0.80)	(0.80)	(1.07)
Electricity	28	27	33	88	30	32	4	66	154
	(22.40)	(21.77)	(26.19)	(23.47)	(24.00)	(25.81)	(3.20)	(17.65)	(20.56)
Kerosene	91	91	93	275	91	89	117	297	572
	(73.39)	(73.39)	(73.81)	(73.33)	(72.80)	(71.77)	(93.60)	(79.41)	(76.37)
Both Electricity and Kerosene	0	1	0	1	3	2	0	5	6
	(0.00)	(0.81)	(0.00)	(0.27)	(2.40)	(1.61)	(0.00)	(1.34)	(0.80)
Other	2	1	0	3	0	0	3	3	6
	(1.60)	(0.81)	(0.00)	(0.80)	(0.00)	(0.00)	(2.40)	(0.80)	(0.80)
Not Reported	1	2	0	3	0	0	0	0	3
	(0.80)	(1.61)	(0.00)	(0.80)	(0.00)	(0.00)	(0.00)	(0.00)	(0.40)
Total	125	124	126	375	125	124	125	374	749
	(100)	(100)	(100)	(100)	(100)	(100)	(100)	(100)	(100)

TABLE 5.20
Main Sources of Fuel

Sources	Inside the Gardens				Inside the Gardens			
	Districts			Total	Districts			Total
	1	2	3		1	2	3	
LPG	4	9	0	13	15	14	2	31
	(3.2)	(7.3)	(0.0)	(3.5)	(12.0)	(11.3)	(1.6)	(8.3)
Electricity	0	0	0	0	0	0	1	1
Kerosene	0	1	1	2	0	0	1	2
	(0.0)	(0.8)	(0.8)	(0.5)	(0.0)	(0.0)	(0.8)	(0.5)
Firewood	120	114	125	359	108	109	118	335
	(96.0)	(91.9)	(99.2)	(95.7)	(86.4)	(87.9)	(94.4)	(89.6)
Cow dung Cakes	0	0	0	0	0	0	1	1
	(0.0)	(0.0)	(0.0)	(0.0)	(0.0)	(0.0)	(0.8)	(0.3)
Others	1	0	0	1	1	1	2	3
	(0.8)	(0.0)	(0.0)	(0.3)	(0.8)	(0.8)	(1.6)	(1.1)
Total	125	124	126	375	125	124	125	374
	(100.0)	(100.0)	(100.0)	(100.0)	(100.0)	(100.0)	(100.0)	(100.0)

while conditions of living of the workers and their families within the gardens is far from comfortable, those moving out of the gardens also find it difficult to access the basic amenities for survival.

Conclusion

Tea garden labourers facing the prospect of retrenchment and unemployment try to find livelihood opportunities outside the garden. But such an effort cannot possibly be adequately described through an individual utility maximising model. The decisions in such cases are typically taken at the household level, considering the livelihood assets at the disposal of the household. An attempt was made to find out the perceptions of the workers of different categories regarding occupational change and diversification. The factors encouraging or discouraging the workers to move out of the tea sector, discussed on the basis of qualitative and quantitative information collected through the primary survey, suggest that a number of different issues, operating at different levels, enter into the making of the decision to move out. The overall findings from the perception analysis, notwithstanding its limitations, suggest that tea garden permanent workers have relatively less inclination to change their occupations. On the other hand, those working as casual workers in tea gardens have expressed greater willingness to shift their occupation. A multiplicity of factors might have played a role in shaping these perceptions. Several factors have acted as the possible determinants of such a pattern of occupational change and mobility.

One of the key constraints faced by those who want to move into other occupations is lack of education and skills. But the tea garden labourers and their children face a number of hurdles in getting access to better education. The mean level of education among the workers was too low for securing access to better employment opportunities outside or inside the garden. Inadequate access to capital is one of the most widely cited obstacles to growth of entrepreneurship and investment. Poor road connectivity and other infrastructural bottlenecks might also have played a role. Another significant factor influencing the occupational diversification process is the nature of the local economy. Also, human security aspects, such as fear of social exclusion and discrimination along ethnic lines play a role in constraining the choice made by some households (Mishra et al. 2011). Finally, the relatively better social security provisions in the gardens provide an important incentive for the workers to continue with the tea garden occupation rather than moving out. An analysis of the living conditions within and outside the gardens brings out the constraints faced by labour households when they move out of the gardens. This also explains why labour households crowd-in the garden labour force rather than joining the labour market elsewhere.

6

Conclusions

The ongoing processes of globalisation and liberalisation have brought about many fundamental changes in developing economies. They have opened up new opportunities for many sectors and industries, while forcing many others to close down or adopt sweeping changes in their production organisation. The Indian economy has entered a higher growth trajectory in the past two decades, and more particularly since 2003–4, but with significant divergences in the growth experiences across regions, states and sectors.

India's northeastern region has been one of the country's least developed regions. Despite its rich natural resource base and outstanding performance in terms of a few human development indicators (such as in literacy rates in few states), states in this region have lagged behind many other states of the country. The region has also experienced prolonged phases of political instability, insurgencies and violence. Special economic packages and transfers from the central government have been used as instruments for economic development of the region. However, weak governance mechanisms and institutional failures, among others, have severely curtailed the development impact of such financial interventions. In the post-liberalisation phase, as India's federal polity is gradually responding to the challenges of neo-liberalism and global economic integration, the states are increasingly competing against each other to attract private capital, both domestic and foreign. The northeastern region is being projected as a gateway to the East and under India's 'Look East Policy'. This policy aims at strengthening India's economic and strategic ties with the Southeast and East Asian economies. India's northeast is expected to benefit from these new and emerging economic linkages. Efforts have been made to develop the communication networks in the region as well. However, because of its geographical isolation, political instability and poor infrastructure the region has so far not seen large inflows of private capital. In this backdrop, the performance of the already existing industries has a critical implication for the states of the region. Assam's tea industry, along with oil and coal, has been at the centre of the state's economy. The performance of this sector, thus, has a crucial bearing upon the economic performance of Assam.

Historically, the tea sector has played an important role in the regional economy of Assam and northeast India. There has been a phenomenal growth of tea gardens in India, particularly since the 1990s, mainly due to increase in the number of small tea gardens in various states such as Assam, Himachal Pradesh, Tamil Nadu and Kerala. This has also altered the regional distribution and concentration of tea gardens. While the south Indian states accounted for nearly 80 per cent of tea gardens in the early 1980s, in recent years their share has come down to 47 per cent. After the inclusion of small tea gardens in the data on total tea gardens since 1998, the share of Assam has jumped from less than 7 per cent to 30–35 per cent in recent years.

The tea sector has been undergoing a serious crisis since the early 1990s. The analysis of the growth performance and structural features of the tea sector in Assam at a disaggregated level brings out the various dimensions of the current crisis in the tea industry. There has been a significant expansion of small tea gardens in Assam, particularly in last two decades. But in terms of share in output their contribution remains marginal. Notwithstanding the growth in area under tea because of increase in the number of small tea gardens, in terms of growth performance in production and yield of tea, the overall scenario remains bleak. The growth performance of the tea sector in recent years has worsened. Assam has been at the centre stage of the worsening performance of the sector. Assam is not an exception to the crisis; in almost every aspect of its manifestation, it has followed the national trend. Our analysis at the district level points out that the poor performance of the tea industry could be because of the failure in raising productivity at an appreciable rate. Although the crisis per se was not the focus of the analysis, our findings on the changing growth performance on the productivity front demonstrates the pervasiveness and profundity of the problems faced by the industry.

Further, if the crisis is one of low and declining productivity, unmistakably it is a crisis faced, more than anywhere else, in the large and medium-sized gardens of the state. No doubt the entry of a large number of small tea gardens has radically altered the structure of the tea industry in the country, and in Assam in particular, but as of now, the core of the solution has to be found in the old, large and medium-sized gardens. Because of their share in total production, and more importantly the high negative growth rates in yield in some of these gardens in the recent years, they have contributed significantly to the spirals of low replanting and low yield. Changes in the age distribution of bushes at the district level seem to suggest that although the share of area under older, relatively less productive bushes has declined at the state level between 1980 to 2002, the share of such bushes continues to remain high in those districts where there has been relatively less significant growth in the area under small gardens. This seems to suggest that older and larger plantations have not been investing enough to plant new bushes. Needless to add, all these changes have profound implications for the labour depending on the tea sector for their livelihoods.

The plantation economies in general, and the tea sector in particular, historically, have been associated with bondage and indenture labour systems. It is true that state intervention in the post-independence period, as well as restructuring of the production process, have brought out several changes in the organisation of production within the gardens. This is again one of the sectors where rates of unionisation are very high. However, there seems to be limited mobility of tea garden workers (or ex-tea garden workers) or their descendants in terms of diversification of sources of earnings and employment. There is considerable degree of 'crowding in' of tea garden workers and their families in the tea sector itself. In the context of rapid and increasing mobility of workers across sectors and spaces, the apparent intergenerational immobility of tea garden labourers is an important, and, to some extent, intriguing question that deserves detailed research attention.

The employment conditions in the tea labour market in Assam, particularly in relation to the changes in employment growth and labour productivity, brings out the problems faced by the tea sector on the employment front. For the period 1980–2004 employment has grown at a higher rate at the all-India level than in Assam. At the all-India level, growth of labour employed has been relatively strong during 1998–2000. This has been possible because of the tremendous expansion of employment in Tamil Nadu; in Assam, however, the growth rate has decelerated significantly. During 1998 and 2004, there has been some recovery of growth of labour per hectare at the all-India level, but the situation in Assam and West Bengal continues to be bleak. As many as four out of the seven districts in Assam under consideration registered a negative growth in labour per hectare, while others experienced very low growth. Although labour productivity had increased relatively comfortably during the 1980s in Assam, during the 1990s labour productivity growth decelerated in many of the districts of Assam. During the last six years, that is, 1998 to 2004, there has been a substantial deterioration in labour productivity in almost all the major tea-producing states of India. Employment elasticity decreased in Assam in the 1990s, in comparison to the 1980s. At the district level, it has declined for almost all the districts during the 1990s. In fact, employment elasticity was negative in Assam during 1998 to 2004.

With increasing consumer awareness and monitoring of the social and environmental conditions under which production is taking place, there has been a remarkable decline in the share of child labour employed by the tea industry, as per the official data. In Assam, the percentage of child and adolescent workers has declined substantially from 14.12 per cent in 1980 to just around 5 per cent in 2004. But the dependence on child and adolescent labour continues to be high in Assam in comparison to the gardens of West Bengal and south India. Another perceptible feature is that between 1991 and 2004, as a result of the decline in the share of child and adolescent labour, the share of both male and female adult labourers went up in Assam, but the increase in the share of male labourers is far more noteworthy than that of the female labourers. This could be because of the fact that with shrinking employment opportunities female casual labourers are facing difficulty in finding work in the gardens. All these features of the labour market capture the implication of the crises of low productivity in the tea sector for labour households in Assam. Increasingly they find it difficult to get absorbed in the tea sector, their traditional source of livelihood.

The specificity of the labour market scenario in the tea gardens lies in the fact that the supply side is governed by a set of factors that import the characteristics of a reserve army of labour. Descendents of indentured labourers, who have not been well integrated into the local economy, live within the gardens and compete with each other for casual employment within the gardens. In contrast, the demand side of the labour market is highly responsive to the market conditions. With increasing pressure to cut down labour cost, in the face of competition in both the domestic as well as in the international market, tea gardens can retrench labour relatively easily. The market fluctuations in labour demand conditions, our analysis shows, have largely been absorbed by one segment of the labour force, the casual labourers.

The pattern of occupational diversification among adults belonging to tea garden worker households and ex-tea garden labour households brings out the relative significance of different livelihoods sources succinctly. First, the distinctions between 'tea garden labour households' and 'ex-tea garden labour households' seems to be of little analytical significance so far as the sources of livelihood and employment are analysed. Among the tea garden labour households, a substantial portion of adults seek and get work in the tea gardens, but many of them, while living inside the tea gardens depend on a variety of other sources for their livelihoods. Similarly, those who have left the tea garden jobs, for a variety of reasons, have not completely severed their relation with the tea gardens. While the main breadwinners have moved out of the gardens to other occupations, either out of compulsion or choice, other members of the households continue to depend on the gardens for their employment. Many of those who work in the gardens are employed as casual workers, mainly in peak plucking season. Second, there is a great deal of occupational concentration in the tea gardens. There is some occupational diversification among the labour households, but the overwhelming significance of tea gardens as a source of livelihood can hardly be overstressed. The outside jobs are less in the nature of stable, permanent and better earning jobs like government service or private salaried service, most of it is within agriculture or casual work outside agriculture. Within the tea gardens there seems to be very insignificant presence of members of the labour households in the upper strata of jobs, such as those at the managerial, clerical and supervisory level. Opportunities for vertical mobility within the gardens appear to be extremely limited. There was hardly any significance of any other occupation in the previous generation. Another significant aspect is that this seems to be a consistent pattern in all the districts.

The intergenerational occupational mobility matrix created on the basis of data generated from the field survey brings out several key aspects of the occupational change of workers in the tea garden. Among all the occupational categories considered in the survey the highest intergenerational concentration (or immobility) of occupations has been in the categories of permanent and casual labourers in the tea gardens. It is found that among the tea labour households (living within the tea gardens), nearly 65 per cent of those whose fathers were working in the gardens have found a job within the gardens, while 54 per cent of those whose fathers were not in the gardens also work in the gardens. Among those who belong to 'ex-tea garden labour households' and live outside the gardens, the scenario is somewhat different. Nearly 15 per cent of those whose fathers had worked in the gardens were employed in the gardens themselves, while 18 per cent of those whose fathers were working outside the gardens have joined the gardens. A majority of workers of course were engaged either as cultivators or as agricultural labourers.

It is found that among the tea labour households the significant determinants of occupational diversification are age, gender and educational qualification of the worker and the nearness of the garden to an urban location. While a relatively higher age decreases the probability of entering a non-tea labour occupation, being a female increase the chances of being in the non-tea labour occupation. An education of up to eight years of schooling increases the possibility of being absorbed outside the tea garden significantly. Nearness to an urban location has a positive impact on occupational diversification at the individual level. On the other hand the determinants of occupational diversification among the individuals in the households living outside the gardens seem to be somewhat different. The significant variables are age, gender and nearness to urban location.

While with higher age the probability of entering non-tea occupation increases, being a male also increases the probability of working in a non-tea occupation.

An attempt was made to find out the perceptions of the workers of different categories regarding occupational change and diversification. The factors encouraging or discouraging the workers to move out of the tea sector was also discussed on the basis of qualitative and quantitative information collected through the primary survey. The overall findings from the perception analysis, notwithstanding its limitations, suggest that tea garden permanent workers have relatively less inclination to change their occupations. On the other hand, those working as casual workers in tea gardens have expressed greater willingness to shift their occupation. A multiplicity of factors might have played a role in shaping these perceptions. Several factors have acted as the possible determinants of such a pattern of occupational change and mobility.

One of the key constraints faced by those who want to move into other occupations is lack of education and skills. But the tea garden labourers and their children face a number of hurdles in getting access to better education. The mean level of education among the workers was too low for securing access to better employment opportunities outside or inside the garden. Inadequate access to capital is one of the most widely cited obstacles to growth of entrepreneurship and investment. Poor connectivity and other infrastructural bottlenecks might also have played a role. Another significant factor influencing the occupational diversification process is the nature of the local economy. Also, human security aspects such as fear of social exclusion and discrimination along ethnic lines do play a role in constraining the choice of some households. Finally, the relatively better social security provisions in the gardens provide an important incentive for workers to continue with the tea garden occupation rather than moving out.

The production and employment conditions in Assam's tea gardens reflect the pressures of structural changes that have been part of the post-reform restructuring of the Indian economy. While there is increasing emphasis on profitability, efficiency and technological upgradation at almost all levels, the burdens of these changes are being shared unequally. In specific regions and sectors this has also meant the strengthening of entrenched structures of exclusion, discrimination and exploitation. The specific case of tea garden labourers that has been investigated here, notwithstanding the very specific nature of the problem, is not an exception so far as the broader changes in the world of work in India are concerned. The reshaping of the economy under neo-liberal reforms has unleashed new entrepreneurial energy and India's recent growth performance has been quite remarkable in comparison with her past. Yet, at the same time, the terms and conditions under which labour is being incorporated into the globally integrated circuits of production, consumption and exchange need to be taken into account while judging the implications of this economic growth for a majority of India's poor. Assam's tea garden workers, like their forefathers who were brought as indentured labourers and were forced to work under subhuman conditions, continue to produce for a global market. Unlike their forefathers, they can, theoretically at least, claim the protection of a democratic constitution. They have been participating in elections, even electing some members of their own 'community' to the legislative assembly as well as to the parliament of India. Their voice is also represented through trade unions, civil society organisations, media and independent citizens. Still, a substantial majority of them have not been able to look beyond the gardens for their livelihood. Those among the 'tea labour community' who have stepped out of the gardens, either out of choice or out of compulsion, have been looking for livelihood options in low earning occupations. The 'invisible chains', that seem to bind them with the gardens, operate at multiple levels limiting their capabilities to search for alternative sources of livelihoods.

Appendix Ia

The Plantations Labour Act, 1951

[Act 69 of 1951]

with

Labour Laws (Exemption from Furnishing Returns and
Maintaining Registers by Certain Establishments Act, 1988

and

SHORT NOTES

Plantations Labour Act, 1951

[Act 69 of 1951 as amended by Acts 42 of 1953, 34 of I960,[1] 53 of
1961, 58 of 1981[2] and 61 of 1986]

[2nd November, 1951]

*An Act to provide for the welfare of labour, and to regulate the conditions of
works, in plantations*

Be it enacted by Parliament as follows:

Prefatory Notes - The following Statement of Objects and Reasons shows
the necessity for this Act:

1. In spite of the fact that the plantation industry provides employment
    for more than a million workers, there is at present no comprehensive
    legislation regulating the conditions of labour in the industry. The
    Tea Districts Emigrant Labour Act, 1932, which applies only to
    Assam regulates merely the conditions for recruitment of labour for
    employment in the tea gardens of Assam. The Workmen's Com-
    pensation Act, 1923, which applies to estates growing cinchona, coffee,
    rubber or tea also does not confer any substantial benefit on planta-
    tion labour as accidents in plantations are few. The other Labour
    Acts like the payment of Wages Act, 1936. Industrial employment
    (Standing Orders) Act, 1946 and Industrial Disputes Act, 1947,
    benefit plantation labour only to a very limited extent. In its report
    the Labour Investigation Committee observed 'that as the conditions
    of the life and employment on plantations were different from those
    in other industries it would be very difficult to fit plantation labour in
    the general framework of the Industrial Labour Legislation without
    creating serious anomalies' and recommended a Plantation Labour
    Code covering all plantation areas.
2. The present Bill drafted as an All India measure seeks to regulate
    the conditions of plantation labour generally. It applies in the first
    instance to tea, coffee, rubber and cinchona plantations, but the state
    Government may apply it to any other plantation. Provision is made
    in the Bill for assuring to the worker reasonable amenities, as for
    example, the supply of wholesome drinking water or suitable medical
    and educational facilities or provisions for canteens and creches in
    suitable cases, or provision for sufficient number of latrines and
    urinals separately for males and females. Housing accommodation is
    also to be provided for every worker and standards and specifications
    of such housing accommodation will be prescribed after due

consultation. The Bill also regulates the working hours of workers employed in plantation.

3. Children under twelve are prohibited for employment in any plantation and State Governments are empowered to make rules regulating payment of sickness or maternity benefits.

4. Necessary provision is made in the Bill for the appointment of a suitable inspecting, medical or other staff for the purposes of securing in plantations of the various provisions in the Act.

Chapter I

Preliminary

1. Short title, extent, commencement and application-

(1) This Act may be called the Plantations Labour Act, 1951.

(2) Itextends to the whole of India except the State of Jammu and Kashmir.

(3) It shall come into force on such as the Central Government may by notification in the Official Gazette, appoint.

[3][(4) It applies to the following plantations, that is to say,-

    (a) to any land used or intended to be used for growing tea, coffee, rubber, [cinchona or cardamom][1] which admeasures[5] [5] hectares or more and In which [fifteen][6] or more persons are employed or were employed on any day of the preceding twelve months;

---

1. The amending Act 34 of 1960, came into force w.e.f. 21-11-1960, vide S.O. 2788, dt. 15–11–1960, published in Gaz. of India, Pt, II, S. 3(ii), dt. 19–11–1960, p. 3352.
2. The amending Act 58 of 1981 came into force w.e.f. 26-1-1982, vide S.O. 45(E), dt. 23-1-1982 (1982-CCL-III-18).
3. Subs, by Act 34 of 1960.
4. Subs, by Act 58 of 1981. (w.e.f. 26-1-1982).
5. Subs, by Act 58 of 1981. (w.e.f. 26-1-1982).

    (b) to any land used or intended to be used for growing any other plant, which admeasures[7][5] hectares or more and in which [8][fifteen]or more persons are employed or were employed on any day of the preceding twelve months, if, after obtaining the approval of the Central Government the State Government, by notification in the Official Gazette, so directs.

[9][Explanation.- Where any piece of land used for growing any plant referred to in clause (a) or clause (b) of this sub-section admeasures less than 5 hectares and is contiguous to any other piece of land not being so used, but capable of being so used, and both such pieces of land are under the management of the same employer, then, for the purpose of this sub-section, the piece of land first mentioned shall be deemed to be plantation, if the total area of both such pieces of land admeasures 5 hectares or more.]

(5) The State Government may, by notification in the Official Gazette, declare that all or any of the provisions of this Act shall apply also to any land used or intended to be used for growing any plant referred to in clause (a) or clause (b) of sub-section (4), notwithstanding that-

(a) It admeasures less than[10] [5] hectares, or

(b) The number of persons employed therein is less than [fifteen][11]:

Provided that no such declaration shall be made in respect of such land which admeasured less than [12] [5] hectares or in which less than[13] [fifteen] persons where employed, immediately before the commencement of this Act.]

Note I.—The Act came into force w.e.f. 1st April, 1954.[14] Extended to Pondicherry by Reg.7 of 1963. Applied to Naga Hills District by Assam Government Noti. Dated 17–11–1954.

Note II.—For exemption from application of this Act to small and very small establishments, see Labour Laws (Exemption from Furnishing Returns and Maintaining Registers by Certain Establishments) Act, 1988 (51 of 1988).

The power conferred under Section 1(5) to the state Government does not amount to excessive delegation of legislative function. K. T. Thomas v. Union of India, (1979) 2 LIJ 290 (Ker.)

2. Definitions.- In this Act, unless the context otherwise requires,

(a) 'adolescent' means a person who has completed his[15] [fourteenth] year but has not completed his eighteenth year,

(b) 'adult' means a person who has completed his eighteenth year,

(c) 'Child' means a person who has completed his[15] [fourteenth] year,

(d) 'day' means a period of twenty-four hours beginning at midnight;

(e) 'employer' when used in relation to a plantation, means the person who has the ultimate control over the affairs of the plantation, and where the affairs of any plantation are entrusted to any other person (whether called a managing agent, manager, superintendent or by any other name) such other person shall be deemed to be the employer in relation to that plantation;

6.	Subs, by Act 58 of 1981 (w.e.f. 26-1-1982)
7.	Subs, by Act 58 of 1981 (w.e.f. 26-1-1982)
8.	Subs, by Act 58 of 1981 (w.e.f. 26-1-1982)
9.	Ins. by Act 58 of 1981 (w.e.f. 26-1-1982)
10.	Subs, by Act 58 of 1981 (w.e.f. 26-1-1982)
11.	Subs, by Act 58 of 1981 (w.e.f. 26-1-1982)
12.	Subs, by Act 58 of 1981 (w.e.f. 26-1-1982)
13.	Subs, by Act 58 of 1981 (w.e.f. 26-1-1982)
14.	See. S.R.O. 880, dt. 6th Mar., 1954, Gaz. of India, 1954, Pt. II, S. 3, p. 530.
15.	Subs, by Act 61 of 1986.

[16][(ee) 'family' , when used in relation to a worker, means -

(i) his or her spouse, and

(ii) the legitimate and adopted children of the worker dependent upon him or, who have not completed their eighteenth year, and includes, where the worker is a male, his parents dependent upon him;]

[17][(eee) 'inspector' means an inspector of plantation appointed under sub-section (1) of section 4 and includes an additional inspector of plantations appointed under sub-section (1-A) of that section;]

[18][(f) 'plantation' means any plantation to which this Act, whether wholly or in part, applies and includes offices, hospitals, dispensaries, schools, and any other premises used for any purpose connected with such plantation, but does not include any factory on the premises to which the provisions of the Factories Act, 1948 (63 of 1948), apply];

(g) 'prescribed' prescribed means prescribed by rules under this Act;

[19][(h) 'qualified medical practitioner' means a person holding a qualification granted by an authority specified or notified under section 3 of the Indian Medical Degrees Act, 1916 (7 of 1916), or specified in the Schedules to the Indian Medical Council Act, 1956 (102 of 1956), and include any person having a certificate granted under any Provincial or State Medical Council Act;

(i) 'Wages' has the meaning assigned to in clause (h) of section 2 of the Minimum Wages Act, 1948 (11 of 1948);

(j) 'week' means a period of seven days beginning at midnight on Saturday night or such other night as may be fixed by the State Government in relation to plantation in any area after such consultation as may be prescribed with reference to the plantations concerned in that area;

[20][(k) 'worker' means a person employed in a plantation for hire or reward, whether directly or through any agency, to do any work, skilled, unskilled, manual or clerical, but does not include -

  (i)   a medical officer employed in the plantation
  (ii) any person employed in the plantation (including any member of the medical staff) whose monthly wages exceed [rupees seven hundred and fifty][21];
  (iii) any person employed in the plantation primarily in a managerial capacity, notwithstanding that his monthly wages do not exceed [rupees seven hundred and fifty][21];

---

16. ins. by Act 34 of 1960.
17. ins. by Act 58 of 1981 (w.e.f. 26-1-1982).
18. subs, by Act 34 of 1960.
19. subs, by Act 34 of 1960.
20. subs, by Act 34 of 1960.
21. subs, by Act 58 of 1981. (w.e.f. 26-1-1982).

(iv) any person temporarily employed in the plantation in any work relating to the construction, development or maintenance of buildings, roads, bridges, drains or canals;]

(1) 'young person' means a person who is either a child or an adolescent.

3. Reference to time of day :- In this Act, references to time of day are references to Indian Standard Time being five and a half hours ahead of Greenwich Mean Time:

Provided that for any area in which the Indian Standard Time is not ordinarily observed, the State Government may make rules-

  (a) specifying the area;
  (b) defining the local mean time ordinarily, observed therein; and
  (c) permitting such time to be observed in all or any of the plantations situated in that area.

[22][Chapter I-A

Registration of Plantations

3-A. Appointment of registering officers: - The state Government may, by notification in the Official Gazette-

(a) appoint such persons, being Gazette Officers of Government as it thinks fit, to be registering officers for the purposes of this Chapter, and

(b) define the limits within which a registering officer shall exercise the powers and discharge the functions conferred or imposed on him by or under this chapter.

3-B. Registration of plantations: -

(1) Every employer of a plantation existing at the commencement of the Plantations Labour (Amendment) Act, 1981 shall within a period of sixty days of such commencement, and every employer of any other plantation coming into existence after such commencement shall, within a period of sixty days of the coming into existence of such plantation, make an application to the registering officer for the registration of such plantation:

Provided that the registering officer may entertain any such application after the expiry of the period aforesaid if he is satisfied that the applicant was prevented by sufficient cause from making the application within such period.

(2) Every application made under sub-section (1) shall be in such form and shall contain such particulars and shall be accompanied by such fees as may be prescribed.

(3) After the receipt of an application under sub-section (1), the registering officer shall register the plantation.

(4) Where a plantation is registered under this section, the registering officer shall issue a certificate of registration to the employer thereof in such form as may be prescribed.

(5) Where, after the registration of a plantation under this section, any change occurs in the ownership or management or in the extent of the area or other prescribed particulars in respect of such plantation, the particulars regarding such change shall be intimated by the employer to the registering officer within thirty days of such change in such form as may be prescribed.

> 22. Chap. I-A ins. By Act. 58 of 1981 (w.e.f. 26-1-1982).

(6) Where as a result of any intimation received under sub-section (5), the registering officer is satisfied that the plantation is no longer required to be registered under this section, he shall, by order in writing, cancel the registration thereof and shall, as soon as practicable, cause such order to be published in any one newspaper in the language of, and having circulation in, the area where the plantation is situated.

3-C. Appeals against orders of registering officer-

(1) Any person aggrieved by the order of a registering officer under sub-section (6) of Section 3-B may, within thirty days of the publication of such order in the newspaper under that sub-section, prefer and appeal to such authority as may be prescribed.

Provided that the appellate authority may entertain an appeal under this sub-section after the expiry of the aforesaid period if it is satisfied that the appellant was prevented by sufficient cause from preferring the appeal within such period.

(2) After the receipt of an appeal under sub-section(l), the appellate authority may, after giving the appellant, the employer referred to in sub-section (5) of Section 3-B and the registering officer an opportunity of being heard in the matter, dispose of the appeal as expeditiously as possible.

Notes

Principles for condoning delay:- In dealing with the question of condoning the delay under Section 5 of the Limitation Act the party seeking relief has to satisfy the Court that it had sufficient cause for not preferring the appeal or making the application within the prescribed time and this has always been understood to mean that the explanation has to cover the whole period of delay. It is not possible to lay down precisely as to what facts or matters would constitute sufficient cause under the section but those words should be liberally construed so as to advance substantial justice where no negligence or any inaction or want of bona fides is imputable to the party; that is, the delay in filing the appeal should not have been for reasons which indicated the party's negligence is not taking necessary steps which he would have or should have been taken. Discretion is conferred on the High Court before which and application for condoning the delay is made and if the Court after keeping in view relevant principles exercises its discretion grant in relief, unless it is shown to be manifestly unjust or perverse, the Supreme Court would be loathe to interfere with it. Sandhay Rani Sarkar v V. Sudha Rani Devi, (1978) 2 SCC 116: AIR 1978 SC 537.

Section 5 of the Limitation Act is a hard task-master and judicial interpretation has encased it within a narrow compass. A large measure of case-law has grown around Section 5, its highlights being that one ought not easily to take away a right which has accrued to a party by lapse of time and that therefore a litigant who is not vigilant about his right must explain every day's delay. These and similar considerations which influence the decision of Section 5 applications are out of place in cases where the appeal itself is preferred within the period of limitation but there is an irregularity in presenting it. Bikram Dass Chela v. Financial Commissioner, Revenue, (1977) 4 SCC 69; AIR 1977 SC 2221.

3-D. Power to make rules.-

(1) The State Government may, by notification in the Official Gazette, make rules to carry out the purposes of this Chapter.
(2) In particular, and without prejudice to the generality of the foregoing power, such rules may provide for all or any of the following matters, namely:-

- (a) the form of application for the registration of a plantation, the particulars to be contained in such application and the fees to be accompanied along with such application;
- (b) the form of the certificate of registration;
- (c) the particulars regarding any change in respect of which intimation shall be given by the employer to the registering officer under sub-section (5) of Section 3-B and the form in which such change shall be intimated;
- (d) the authority to which an appeal may be preferred under Section 3-C and the fees payable in respect of such appeal;
- (e) the registers to be kept and maintained by a registering officer.

Chapter II

Inspecting Staff

4. Chief inspector and inspectors.- (1) The State Government may, by notification in the Official Gazette, appoint for the State a duly qualified person to be the chief inspector of plantation and so may duly qualified persons to be inspectors of plantations subordinate to the chief inspector as it thinks fit.

[23][(1-A) The State Government may also, by notification in the Official Gazette, appoint such officers of the State Government or of any local authority under its control as it thinks fit, to be additional inspectors of plantations for all or any of the purposes of this Act.]

(2) Subject to such rules as may be made in this behalf by the State Government, the chief inspector may declare the local area or areas within which, or the plantations with respect to which, inspectors shall exercise their powers under this Act, and may himself exercise the powers of an inspector within such limits as may be assigned to him by the State Government.

(3) The chief inspector and all inspectors shall be deemed to be public servants within the meaning of the Indian Penal Code (45 of 1860).

23. Ins. By Act 58 of 1981 (w.e.f. 26-1-1982)

Note - This section empowers the State Government to appoint duly qualified Inspecting Staff with a Chief Inspector of Plantations and many inspectors under hid.

5. Powers and functions of inspectors- Subject to any rules made by the State Government in this behalf, an inspector may within the local limits for which he is appointed-

> (a) make such examination and inquiry as he thinks fit in order to ascertain whether the provisions of this Act and of the rules made thereunder are being observed in the case of any plantation;
>
> (b) with such assistants, if any, as he thinks fit, enter, inspect and examine any plantation or part thereof at any reasonable time for the purpose of carrying out the objects of this Act;
>
> (c) examine the crops grown in any plantation or any worker employed therein or require the production of any register or other document maintained in pursuance of this Act, and take on the spot or otherwise statements on any person which he may consider necessary for carrying out the purposes of this Act;
>
> (d) exercise such other powers as may be prescribed:
> provided that no person shall be compelled under this section to answer any question or make any statement tending to incriminate himself.

6. Facilities to be afforded to inspectors- Every employer shall afford the inspector all reasonable facilities for making any entry, inspection, examination or enquiry under this Act.

7. Certifying surgeons- (1) The State Government may appoint qualified medical practitioners to be certifying surgeons for the purposes of this Act within such local limits or for such plantation or class of plantations as it may assign to them respectively.

(2) The certifying surgeon shall carry out such duties as may be prescribed in connection with-

> (a) the examination and certification of workers;
>
> (b) the exercise of such medical supervision as may be prescribed where adolescents and children are, or are to be, employed in any work in any plantation which is likely to cause injury to their health.

Chapter III

Provisions As To Health

8. Drinking water - In every plantation effective arrangements shall be made by the employer to provide and maintain at convenient places in the plantation a sufficient supply of wholesome drinking water for all workers.

9. Conservancy- (1) There shall be provided separately for males and females in every plantation a sufficient number of latrines and urinals of prescribed types so situated as to be convenient and accessible to workers employed therein.

(2) All latrines and urinals provided under sub-section (1) shall be maintained in a clean and sanitary condition.

10. Medical facilities- (1) In every plantation there shall be provided and maintained so as to be readily available such medical facilities for the workers [and their families]² as may be prescribed by the State Government.

(2) If in any plantation medical facilities are not provided and maintained as required by sub-section (1) the chief inspector may cause to be provided and maintained therein such medical facilities, and recover the cost thereof from the defaulting employer.

(3) For the purposes of such recovery the chief inspector may certify the costs to be recovered to the collector, who may recover the amount as an arrear of land revenue.

Chapter IV

Welfare

11. Canteens -(1) The State Government may make rules requiring that in every plantation wherein one hundred and fifty workers are ordinarily employed, one or more canteens shall be provided and maintained by the employer for the use of the workers.

(2) Without prejudice to the generality of the foregoing power, such rules may provided for-

(a) the date by which the canteen shall be provided;
(b) the number of canteens that shall be provided and the standards in respect of construction, accommodation, furniture and other equipment of the canteen;
(c) the foodstuffs which may be served therein and the charges which may be made therefore.
(d) the constitution of a managing committee for the canteen and the representation of the management of the canteen;
(e) The delegation to the chief inspector, subject to such conditions as may be prescribed, of the power to make rules under clause (c).

Note- Under this section the State Governments are empowered to ask employers to open canteens in the plantations wherein one hundred and fifty workers are employed and to make rules for the working and maintenance of canteens

24. Ins by Act 34 of 1960

12. Creches- [25][(1) In every plantation wherein fifty or more women workers (including women workers employed by any contractor) are employed or were employed on any day of the preceding twelve months, or where the number of children of women workers (including women workers employed by any contractor) in twenty or more, there shall be provided and maintained by the employer suitable rooms for the use of children of such women workers. Explanation- For the purposes of this sub-section and sub-section (1-A), "children" means persons who are below the age of six years.]

[26][(1-A) Notwithstanding anything contained in sub-section (1), if, in respect of any plantation wherein less than fifty women workers (including women workers employed by any contractor) are employed or were employed on any day of the preceding twelve months, or where the number of children of such women workers is less than twenty, the State Government, having regard to the number of children of such women workers deems it necessary that suitable rooms for the use of such children should be provided and maintained by the employer, it may, by order, direct the employer to provide and maintain such rooms and threupon the employer shall be bound to comply with such direction.]

(2) [27][The rooms referred to in sub-section(l) or sub-section 1-A] shall-

(a) provide adequate accommodation;
(b) be adequately lighted and ventilated;
(c) be maintained in a clean and sanitary condition; and
(d) be under the charge of a woman trained in the care of children and infants.

(3) The State Government may make rules prescribing the location and the standards of [28][the rooms referred to in sub-section(l) or sub-section (1-A) in respect of their construction and accommodation and the equipment and amenities to be provided therein.

13. Recreational facilities- The State Government may make rules requiring every employer to make provision in his plantation for such recreational facilities for the workers and children employed threin as may be prescribed.

14. Educational facilities- Where the children between the ages of six and twelve of workers employed in any plantation exceed twenty-five in number, the State Government may make rules requiring every employer to provide educational facilities for the children in such manner and of such standard as may be prescribed.

25. Subs. By Act 58 of 1981 (w.e.f. 26-1-1982).
26. Subs. By Act 58 of 1981 (w.e.f. 26-1-1982).

27. Subs. By Act 58 of 1981 (w.e.f. 26-1-1982).
28. Subs. By Act 58 of 1981 (w.e.f. 26-1-1982).

[29][**15. Housing facilities**- It shall be the duty of every employer to provide and maintain necessary housing accommodation-

(a) for every worker (including his family) residing in the plantation;

(b) for every worker (including his family) residing outside the plantation, who has put in six months of continues serviced in such plantation and who has expressed a desire in writing to reside in the plantation;

(c) Provided that the requirement of continuous service of six months under this clause shall not apply to a worker who is a member of the family of a deceased worker who, immediately before his death, was residing in the plantation.]

Notes- Although Sections 15 and 16 cast a duty on the management to provide minimum residential accommodation to its workmen, Rule 65(2) of the Assam Plantations Labour Rules, 1956, clearly lays down that the occupant of a house shall not make any unauthorized additions to or alterations in the house. Where a workman started construction a house adjoining his quarter provided by the management without the management's permission and despite the management's repeated orders to dismantle the construction, completed the same and on being charge-sheeted, submitted explanation expressing inability to dismantle the house on the ground that he had installed a deity there, hence the management was justified in proceeding with a domestic enquiry against him and discharging him as a result of the report of the enquiry. Hattilai T. E. v. Presiding Officer, Labour Court, 1976 Lab IC 172 Gau).

16-. Power to make rules relating to housing- The State Government may make rules for the purposes of giving effect to the provisions of Section 15 and, in particular providing for-

(a) the standard and specification of the accommodation to be provided;

(b) the selection and preparation of sites for the construction of houses and the size of such plot;

(c) the constitution of advisory boards consisting of representatives of the State Government, the employer and the workers for consultation in regard to matters connected with housing and the exercise by them of such powers, functions and duties in relation thereto as may be specified;

(d) the fixing of rent, if any, for the housing accommodation provided for workers;

(e) the allotment to workers and their families of housing accommodation and suitable strips of vacant land adjoining such accommodation for the purposes of maintaining kitchen gardens, [* * *]³⁰ and for the eviction of workers and their families from such accommodation;

29. Subs. By Act 58 of 1981 (w.e.f. 26-1-1982).

30. Omitted by Act 34 of 1960.

(f) access to the public to those parts of the plantation wherein in workers are housed.

³¹[**16-A.** Liability of employer in respect of accidents resulting from collapse of houses provided by him- (1) If death or injury is caused to any worker or a member of his family of result of the collapse of a house provided under Section 15, and the collapse is not solely and directly attributable to a fault on the part of any occupant of the house or to natural calamity, the employer shall be liable to pay compensation.

(2) The provisions of Section 4 of, and Schedule IV to, the Workmen's Compensation act, 1923 (8 of 1923), as in force for the time being, regarding the amount of compensation payable to a workman under that Act shall, so far as may be, apply for the determination of the amount of compensation payable under sub-section (1).

16-B. Appointment of Commissioners- The State Government may, by notification in the Official Gazette, appoint as many persons, possessing the prescribed qualifications, as it thinks fit, to be commissioners to determine the amount of compensation payable under Section 16-A and may define the limits within which each such Commissioner shall exercise the powers and discharge the functions conferred or imposed on him by or under this Act.

31. Ss. 16-A to 16-G ins. By Act 58 of 1981 (w.e.f. 26-1-1982).

Compensation is to be awarded under the Workmen's Compensation Act on the basis of Swages' of the injured employee. When the question is what compensation is to be awarded to an employee who has been injured, the term Swages' is to be interpreted in the light of the definition given in Section 2(n) of the Workmen's Compensation act, and not in the light of the definition given in the Payment of Wages Act. The words 'privilege or benefit' in the definition of Wages' in the Workmen's Compensation Act includes the benefit of free accommodation. It is, therefore, clear that the monetary value of such accommodation, since free accommodation is capable of being estimated in money, when provided free to an applicant falls within the term Swages' for the purpose of assessing the amount of compensation. B. M. & G. Engineering Factory v. Bahadur Singh, AIR 1955 All 182 (DB).

16-C. Application for compensation- (1) An application for payment of compensation under Section 16-A may be made to the Commissioner-

    (a) by the person who has sustained the injury; or

    (b) by any agent duly authorized by the person who has sustained the injury; or

    (c) where the person who has sustained the injury is a minor, by his guardian; or

    (d) where death has resulted out of collapse of the house, by any dependant of the deceased or by any agent duly authorized by such dependant or if such dependant is a minor, by this guardian.

(2) Every application under sub-section (1) shall be in such form and shall contain such particulars as may be prescribed.

(3) No application under this section shall be entertained unless it is made within six months of the collapse of the house:

Provided that the Commissioner may, if he is satisfied that the applicant was prevented by sufficient cause from making the application within the aforesaid period of six months, entertain such application within a further period of six months.

Explanation- In this section, the expression "dependant" has the meaning assigned to it in clause (d) of Section 2 of the Workmen's Compensation Act, 1923 (8 of 1923).

16-D. Procedure and powers of Commissioner -

(1) On receipt of an application under Section 16-C, the Commissioner may make an inquiry into the matter covered by the application.

(2) In determining the amount of compensation payable under Section 16-A, the Commissioner may, subject to any rules that may be made in this behalf, follow such summary procedure as he thinks fit.

(3) The Commissioner shall have all the powers of a cilvil court while trying a suit under the Code of Civil Procedure, 1908 (5 of 1908) in respect of the following matters, namely :-

    (a) summoning and enforcing the attendance of any person and examining him on oath;

    (b) requiring the discovery and production of any document;

    (c) receiving evidence on affidavits;

(d) requisitioning any public record or copy thereof from any court or office;

(e) issuing commissioners of the examination of witnesses or documents'

(f) any other matter which may be prescribed.

(4) Subject to any rules that may be made in this behalf, the Commissioner may, for the purpose of determining any claim or compensation, choose one or more persons possessing special knowledge of any matter relevant to the inquiry to assist him in holding the inquiry.

16-E. Liability to pay compensation, etc., to be decided by Commissioner

(1) Any question as to liability of an employer to pay compensation under Section 16-A, or as to the amount thereof, or as to the person to whom such compensation is payable, shall be decided by the Commissioner.

(2) Any person aggrieved by a decision of the Commissioner refusing to grant compensation, or as to the amount of compensation grant to whom, or to the apportionment thereof, may prefer and appeal to High Court having jurisdiction over the place where the collapse of the house had occurred, within ninety days of the communication of the order to the Commissioner to such person:

Provided that the High Court may entertain any such appeal after the expiry of the period aforesaid if it is satisfied that the appellant was prevented by sufficient cause from preferring the appeal within such period:

Provided further that nothing in this sub-section shall be deemed to authorize the High Court to grant compensation in excess of the amount of compensation payable under Section 16-A.

(3) Subject to the decision of the High Court in cases in which an appeal is preferred under sub-section (2), the decision of the Commissioner under sub-ection (1) shall be final and shall not be called in question in any court.]

16-F. Saving as to certain rights- The right of any person to claim compensation under sub-section 16-A shall be without prejudice to the right of such person to recover compensation payable under any other law for the time being in force; but no person shall be entitled to claim compensation more than once in respect of the same collapse of the house.

16-G. Power to make rules - (1) The State Government may, by notification in the Official Gazette, make rules for giving effect to the provisions of Sections 16-A to 16-F (both inclusive). (2) In particular, and without prejudice to the generality of the foregoing power, such rules may provide for-

(i) the qualifications and conditions of service of Commissioners;

(ii) the manner in which claims for compensation may be inquired into and determined by the Commissioner;

(iii) the matters in respect of which any person may be chosen to assist the Commissioner under Section 16-D and the functions that may be performed by such person;

(iv) generally for the effective exercise of any powers conferred on the commissioner.]

17. Other facilities: - The State Government may make rules requiring that in every plantation the employer shall provide the workers with such number and type of umbrellas, blankets, rain coats or other like amenities for the protection of workers from rain or cold as may be prescribed.

18. Welfare officers - (1) In every plantation wherein three hundred or more workers are ordinarily employed the employer shall employ such number of welfare officers as may be prescribed.

(2) The State Government may prescribe the duties, qualifications and conditions of service of officers employed under sub-section (1).

State Amendment

Kerala. - In Section 18, after sub-section (1) add the following : (1-A) if in any plantation, welfare officers are not employed as required by the rules made under sub-section (1), the chief inspector may appoint the required number of welfare officers and there upon such officers shall be deemed to have been employed by the employer under sub-section (1):

Provided that before appointing welfare officers under this sub-section the employer shall be given an opportunity of being heard- Kerala Act 25 of 1969, S. 2 (1-12-1969).

Section 18-a

Kerala- After Section 18, add the following section:
18-A. Chief Inspector to provide facilities on default by employer - (1) If in any plantation, facilities are not provided or maintained by the employer as required by Section 8 or Section 9 or Section 12 or Section 15 or the rules made under Section 11 or Section 14 or section 17, the chief inspector may cause to be provided or maintained therein such facilities and recover the cost thereof from the defaulting employer:

Provided that before providing or maintaining such facilities the employer shall be given an opportunity of being heard.

(2) For the purpose of all recovery of the cost under subsection (1) the chief inspector may certify the amount to be recovered to the Collector, who may thereupon recover such amount as an arrear of land revenue. Kerala Act 25 of 1969, S.2 (1-12-1969).

Chapter V
Hours and Limitation of Employment

19. Weekly hours- [3][(1)] Save as otherwise expressly provided in this Act, no adult worker shall be required or allowed to work on any plantation in excess of [4][forty-eight hours] a week and no adolescent or child for more than [5][twenty-seven hours] a week.

[35][(2) Where an adult worker works in any plantation on any day in excess of the number of hours constituting a normal working day or for more than forty-eight hours in any week, he shall, in respect of such overtime work, be entitled to twice the rates of ordinary wages:

Provided that no such worker shall be allowed to work for more than nine hours on any day and more than fifty-four hours in any week.

(3) For any work done on any closed holiday in the plantation or on any day of rest, a worker shall be entitled to twice the rates of ordinary wages as in the case of overtime work.]

20. Weekly holidays- (1) The State Government may by rules made in this behalf-

> (a) provide for a day of rest in every period of seven days which shall be allowed to all workers;

32 Renumbered by Act 58 of 1981 (w.e.f. 26-1-1982).
33 Subs, by Act 58 of 1981 (w.e.f. 26-1-1982).
34 Subs, by Act 58 of 1981 (w.e.f. 26-1-1982).
35 Subs, by Act 58 of 1981 (w.e.f. 26-1-1982).

[36][(b) provide for the conditions subject to which, and the circumstances in which, and adult worker may be required or allowed to work overtime.]

(2) Notwithstanding anything contained in clause (a) of subsection (1) where a worker is willing to work on any day of rest which is not a closed holiday in the Plantation, nothing contained in this section shall prevent him from doing so:

Provided that in so doing a worker does not work for more than ten days consecutively without a holiday for a whole day intervening.

Explanation 1- Where on any day a worker has been prevented from working in any plantation by reason of tempest, fire, rain or other natural causes, that day, may, if he so desires, be treated as his day of rest for the relevant period of seven days within the meaning of sub-section(l).

Explanation 2- Nothing contained in this section shall apply to any worker whose total period of employment including any day spent on leave is less than six days.

21. Daily intervals for rest- The period of work on each day shall be so fixed that no period shall exceed five hours and that no worker shall work for more than five hours before he has had an interval for rest for at least half an hour.

22. Spread-over- The period of work of an adult worker in a plantation shall be son arranged that inclusive of his interval for rest under Section [37][21] it shall not spread over more than twelve hours including the time spent in waiting for work on any day.

23. Notice of period of work- (1) There shall be displayed and correctly maintained in every plantation a notice of periods of work in such form and manner as may be prescribed showing clearly for every day the periods during which the workers may be required to work.

36. Subs, by Act 58 of 1981 (w.e.f. 26-1-1982).
37. Subs. By Act 42 of 1953, S.4 and Sch. Ill, for : 19".

(2) Subject to the other provisions contained in this Act, no worker shall be required or allowed to work in any plantation otherwise than in accordance with the notice of periods of work displayed in the plantation.

(3) An employer may refuse to employ a worker for any day if on that day returns up for work more than half an hour after the time fixed for the commencement of the day's work.

24. *Prohibition of employment of young children*[38][* * *]

25. Night work for women and children- Except with the permission of the State Government, no woman or child worker shall be employed in any plantation otherwise than between the hours of 6 a.m. and 7 p.m.

Provided that nothing in this section shall be deemed to apply to midwives and nurses employed as such in any plantation.

26. Non-adult workers to carry tokens: - No child [39][* ★ *] and no adolescent shall be required or allowed to work in any plantation unless-

(a) a certificate of fitness granted with the reference to him under Section 27 is in the custody of the employer; and
(b) such child or adolescent carries with him while he is at work a token giving a reference to such certificate.

27. Certificate of fitness- (1) A certifying surgeon shall, no the application of any young person or his parent or guardian accompanied by a document signed by the employer or any other person on his behalf that such person will be employed in the plantation if certified to be fit for work, or on the application of the employer or any other person on his behalf with reference to any young person intending to work, examine such person and ascertain his fitness for work either as a child or as an adolescent.

38. Omitted by Act 61 of 1986.
39. Omitted by Act 61 of 1986.

(2) A certificate of fitness granted under this section shall be valid for a period of twelve months from the date thereof, but may be renewed.

(3) Any fee payable for a certificate under this section shall be paid by the employer and shall not be recoverable from the young person, his parents or guardian.

28. Power to require medical examination- An inspector may, if he thinks necessary so to do, cause any young person employed in a plantation to be examined by a certifying surgeon.

Chapter VI

Leave with Wages

29. Application of chapter - The provisions of this Chapter shall not operate to the prejudice of any rights to which a worker may be entitled under any other law or under the terms of any award, agreement, or contract of service:

Provided that where such award, agreement or contract of service provides for a longer leave with wages than provide in this chapter the worker shall be entitled only to such longer leave.

Explanation - For the purpose of this Chapter leave shall not, except as provided in Section 30, include weekly holidays or holidays for festivals or other similar occasions.

30. Annual leave with wages- (1) Every worker shall be allowed leave with wages for a number of days calculated at the rate of -

    (a) if an adult, one day for every twenty days of works performed by him, and

    (b) if a young person, one day for every fifteen days of work performed by him:

[40] [* * *]

40. Proviso omitted by Act 58 of 1981 (w.e.f. 26–1–982)

[41][*Explanation* [(l)]][42] - For the purposes of calculating leave under this sub-section -

    (a) any day on which no work or less than half a day's work is performed shall not be counted, and

    (b) any day on which half or more than half a day's work is performed shall be counted as one day.]

[43][*Explanation* 2- The leave admissible under this sub-section shall be exclusive of all holidays, whether occurring during, or at either end of, the period of leave.]

(2) If a worker does not in any one period of twelve months take the whole of the leave allowed to him under sub-section (1), and leave on taken by him shall be added to the leave to be allowed to him under that sub-section in the succeeding period of twelve months.

(3) A worker shall cease to earn any leave under this section when the earned leave due to him amounts to thirty days. [44][(4) If the employment of a worker who is entitled to leave under this section is terminated by the employer before he has taken the entire leave to which he is entitled, the employer shall pay him the amount payable under Section 31 in respect of the leave not take, and such payment shall be made before the expiry of the second working day after such termination.]

31. Wages during leave period- [45][(1) For the leave allowed to a worker under Section 30, he shall be paid,-

  (a) if employed wholly on a time-rate basis, at a rate equal to the daily wage payable to him immediately before the commencement of such leave under any law or under the terms of any award, agreement or contract of service, and
  (b) in other cases, including cases where he is, during the preceding twelve calendar months, paid partly on a time-rate basis and partly on a piece-rate basis, at the rate of the average daily wage calculated over the preceding twelve calendar months.

  41. Ins. By Act 34 of 1960.
  42. Ins. By Act 58 of 1981 (w.e.f. 26-1-1982).
  43. Ins. By Act 58 of 1981 (w.e.f. 26-1-1982).
  44. Ins. by Act 34 of 1960.
  45. Subs, by act 34 of 1960.

Explanation- For the purposes of clause (b) of sub-section (1), the average daily wage shall be computed on the basis of this total full-time earnings during the preceding twelve calendar months, exclusive of any overtime earnings of bonus, if any, but inclusive of dearness allowance.

(1-A) In addition to the wages for the leave period at the rates specified in sub-section (1), a worker shall also be paid the cash value of food and other concessions, if any, allowed to him by the employer in addition to his daily wages unless these concessions are continued during the leave period.]
(2) A worker who has been allowed leave for [46][any period not less than] four days in the case of an adult and five days in the case of a young person under Section 30 shall, before his leave begins, be paid his wages for the period of the leave allowed.

⁴⁷32. Sickness and maternity benefits- (1) Subject to any rules that may be made in this behalf, every worker shall be entitled to obtain from his employer -

    (a) in the case of sickness certified by a qualified medical practitioner, sickness allowance, and

    (b) if a woman, in case of confinement or expected confinement, maternity allowance, at such rate, for such period and at such intervals as may be prescribed.

(2) The State Government may make rules regulating the payment of sickness or maternity allowance and any such rules may specify the circumstances in which such allowance shall not be payable or shall cease to be payable, and in framing any rules under this section the State Government shall have due regard to the medical facilities that may be provided by the employer in any plantation.

46. Subs, by Act 42 of 1953, S.4 and Sch. Ill, for: any period less than".

47. On the enforcement of the Maternity Benefit Act, 1961 (53 of 1961) in a State in relation to establishments in that State referred to in S. 1(3) thereof, S. 32 will stand amended as follows :-

    (A) In sub-section(l), the letter and brackets "(a)" before the words "in the case of sickness" the word "and" after the words "sickness allowance" and clause (b) shall be omitted;

    (B) in sub-section (2), the words "or maternity" shall be omitted.

⁴⁸ [Chapter VI-A
Accidents

32-A. Notice of accident: - Where in any plantation, an accident occurs which causes death or which causes any bodily injury to a worker by reason of which the worker injured is prevented from working for a period of forty-eight hours or more immediately following the accident, or which is of such a nature as may be prescribed in this behalf, the employer thereof shall send notice thereof to such authorities, in such form, and within such time, as may be prescribed.

32-B. Register of accidents: - The employer shall maintain a register of all accidents which occur in the plantation in such form and is such manner as may be prescribed.]

Chapter VII
Penalties and Procedure

33. Obstruction: - (1) Whoever obstructs an inspector in the discharge of his duties under this Act or refuses or willfully neglects to afford the inspector any reasonable facility for making any inspection, examination or inquiry authorized by or under this act in relation to any inspection, shall be punishable with imprisonment for a term which may extend to three months, or with fine which may extend to five hundred rupees, or with both.

(2) Whoever willfully refuses to produce on the demand of an inspector any register or other document kept in pursuance of this Act, or prevents or attempts to prevent or does anything which he has reason to believe is likely to prevent any person from appearing before or being examined by an inspector action in pursuance of his duties under this Act, shall be punishable with imprisonment for a term which may extend to three months, or with fine which may extend to five hundred rupees, or with both.

34. Use of false certificate of fitness: - Whoever knowingly uses or attempts to use as a certificate of fitness granted to himself under Section 27 a certificate granted to another person under that section, or having been granted a certificate of fitness to himself, knowingly allows it to be used, or allows and attempt to use it to be made by another person, shall be punishable with imprisonment which may extend to one month, or with fine which may extend to fifty rupees, or with both.

48. Ins. By Act 58 of 1981 (w.e.f. 26-1-1982).

35. Contravention of provision regarding employment of labour: - Whoever, except as otherwise permitted by or under this Act, contravenes any provision of this Act or of any rules made thereunder, prohibiting, restriction or regulation the employment of persons in a plantation, shall be punishable with imprisonment for a term which may extend to three months, or with fine which may extend to five hundred rupees, or with both.

36. Other offences: - Whoever contravenes any of the provisions of this Act or of any rules made thereunder for which no other penalty is elsewhere provided by or under this Act shall be punishable with imprisonment for a term which may extend to three months, or with fine which may extend to five hundred rupees, or with both.

37. Enhanced penalty after previous conviction: - If any person who has been convicted of any offence punishable under this act is again guilty of an offence involving a contravention of the same provision, he shall be punishable on a subsequent conviction with imprisonment which may extend to six months, or with fine which may extend to one thousand rupees, or with both:

Provided that for the purposes of this section no cognizance shall be taken of any conviction made more than two years before the commission of the offence which is being punished.

[49][37-A. power of court of make orders-(l) Where an employer is convicted of an offence punishable under Section 36, the court may, in

addition to awarding any punishment, by order in writing, require him within such period as may be specified in the order (which the court may, if it thinks fit and on an application made in this behalf by the employer, from time to time, extend) to take such measures as may be so specified for remedying the matters in respect of which the offence was committed.

[49]Ins. By Act 58 of 1981 (w.e.f. 26-1-1982).

(2) Where an order is made under sub-section (1), the employer shall not be liable under this Act in respect of the continuation of the offence during the period or extended period, as the case may be, specified by the court, but if, on the expiry of such period or extended period, the order of the court has not been fully complied with, the employer shall be deemed to have committed a further offence and he shall, on conviction, be punishable with imprisonment for a term which may extend to six months and with fine which may extend to three hundred rupees for every day after such expiry.]

38. Exemption of employer from liability in certain cases: -

Where an employer charged with an offence under this Act, alleges that another person is the actual offender, he shall be entitled upon complaint made by him in this behalf to have, on giving to the prosecutor in this behalf three clear days' notice in writing of his intention so to do, that other person brought before the court on the day appointed for the hearing of the case and if, after the commission of the offence has been proved, the employer proves to the satisfaction of the court that-

(a) he has used due diligence to enforce the execution of the relevant provisions of this Act; and

(b) that the other person committed the offence in question without his knowledge, consent or connivance,

the said other person shall be convicted of the offence and shall be liable to the like punishment as if he were the employer and the employer shall be acquitted:

Provided that-

(a) the employer may be examined on oath and this evidence and that of any witness whom he calls in his support shall be subject to cross-examination on behalf of the person he charges to be the actual offender and by the prosecutor, and.

(b) If, in spite of due diligence, the person alleged as the actual offender cannot be brought before the court on the day appointed for the hearing of the case, the court shall adjourn the hearing thereof from time to time so, however, that the total period of such adjournment does not exceed three months, and if, by the end of the said period, the person

Alleged as the actual offender cannot still be brought before the court, the court shall proceed to hear the case against the employer.

39. Cognizance of offences: - No court shall take cognizance of any offence under this Act except on complaint made by, or with the previous sanction in writing of, the chief inspector and no court inferior to that of a Presidency Magistrate or a Magistrate of the second class shall try any offence punishable under this Act.

40. Limitation of prosecutions: - No court shall take cognizance of an offence punishable under this Act unless the complaint thereof has been made or is made within three months from the date on which the alleged commission of the offence came to the knowledge of an inspector:

Provided that where the offence consists of disobeying a written order made by an inspector, complaint thereof may be made within six months of the date on which the offence is alleged to have been committed.

Chapter VIII

Miscellaneous

41. Power to give directions: - The Central Government may give directions to the Government of any State as to the carrying into execution in the State of th3e provisions contained in this Act.

42. Power to exempt: - The State Government may, be order in writing, exempt, subject to such conditions and restrictions as it may think fit to impose, any employer or class of employers from all or any of the provisions of this Act:

Provided that no such exemption [50][other than an exemption from Section 19] shall be granted except with the previous approval of the Central Government.

[50] *Ins* by Act 34 of 1960.

43. General power to make rules: - (1) The State Government may, subject to the condition of previous publication, make rules to carry out the purposes of this Act:

Provided that the date be specified under clause (3) of Section 23 of the General Clauses Act, 1897 (10 of 1897), shall not be less than six weeks from the date on which the draft of the proposed rules was published.

(2) In particular, and without prejudice to the generality of the foregoing power, any such rules may provide for-

    (a) the qualifications required in respect of the chief inspector and inspector;

(b) the powers which may be exercised by inspectors and the areas in which and the manner in which such powers may be exercised;

(c) the medical supervision which may be exercised by certifying surgeons;

(d) the examination be inspectors or other persons of the supply and distribution of drinking water in plantations;

(e) appeals from any order of the chief inspector or inspector and the form in which, the time within which and the authorities to which, such appeals may be preferred;

(f) the time within which housing, recreational, educational or other facilities required by this Act to be provided and maintained may be so provided;

(g) the types of latrines and urinals that should be maintained in plantations;

(h) the medical, recreational facilities that should be provided in plantations;

(i) the form and manner in which notices of period of work shall be displayed and maintained;

(j) the registers which should be maintained by employers and the returns, whether occasional or periodical, as in the opinion of the State Government may be required for the purposes of this Act; [51][* * *]

(k) the hours of work for a normal working day for the purpose of wages and overtime.

[52][(1) any other matter which is required to be, or may be prescribed.]

(3) All rules made under this Act shall, if made by any Government, other than the Central Government, be subject to the previous approval of the Central Government.

LABOUR LAWS (Exemption from Furnishing Returns and Maintaining Registers by Certain Establishments) Act, 1988[1]

[Act 51 of 1988]

[24th September, 1988]

Contents

Sections

3. Amendment of certain labour laws
4. Exemption from returns and registers required under certain labour laws

[51] Omitted by Act 58 of 1981 (w.e.f. 26-1-1982).

[52] Added by Act 58 of 1982 (w.e.f. 26-1-1982)

1. Received the assent of the President on Sept. 24, 1988, and published in the Gaz. of India, Extra, Pt. II, S. 1, dt. Sep. 27, 1988.

5. savings
6. penalty
7. power to amend Form
8. Power to remove difficulties

The Schedule

An Act to provide for the exemption of employers in relation to establishments employing a small number of persons from furnishing returns and maintaining registers under certain labour laws

Be it enacted by Parliament in the Thirty-ninth Year of the Republic of India as follows:

Prefatory Note- Statement of Objects and Reasons: -

t(l) Rules and regulations framed under various labour laws provide for maintenance of registers in prescribed formats and periodical submission of returns in prescribed forms. There has been persistent demands from small business and industrial establishments for the simplification and reduction in the number of forms and registers required to be maintained/submitted by small establishments. A Working Group appointed for the purpose had suggested undertaking of legislation as well as other appropriate measures for simplification and reduction in the number of forms and registers required to be maintained by small establishments. In the light of the recommendations of the Working Group, the matter was considered further at various for and Government has considered it desirable that establishments employing smaller number of persons should be exempted to a limited extend, from maintaining registers in the prescribed form and from submitting the various returns under certain existing labour laws.

(2) A small establishment has been defined in the Bill as an establishment in which not less than ten and not more than nineteen persons are employed or were employed on any day of the preceding twelve months. A very small establishment has been defined as an establishment in which not more than nine persons are employed or were employed on any day of the preceding twelve months.

(3) Small establishments will be required to maintain only three muster registers and will be required to submit only one core return in lieu of the existing returns prescribed under the various labour laws.

Similarly, very small establishments would be allowed to combine the three muster registers into a single

Register, Further, they would be required to submit only one annual core return in lieu of the existing returns prescribed under the various labour laws. The forms of the registers and returns have been prescribed in the Bill itself.

(4) However, in view of the special requirements of social security legislation such as, recovery of contribution from employers and employees, their Accountability, reimbursement, etc, no exemption has been given in relation to social security legislation. The enactments from which exemption is sought to be given have been mentioned in the Schedule to the Bill.

(5) The Bill seeks to achieve the above objects.

1. Short title and commencement - (1) This Act may be called the Labour Laws (Exemption from Furnishing Returns and Maintaining Registers by Certain Establishments) Act, 1988.
2. It extends to the whole of India:
   Provided that nothing contained in this Act, in relation to the Plantations Labour Act, 1951 (69 of 1951) shall extend to the State of Jammu and Kashmir.
3. It shall come into force on such date as the Central Government may, by notification in the Official Gazette, appoint, and different dates may be appointed for different States, and any reference in any provision of this Act to the commencement of this Act shall be construed as a reference to the coming into force of that provision in that State.

Date of Enforcement- The Act came into force in the whole of India w.e.f. 1-5-1989 *vide* Noti. No. G.S.R. 436E) dated 10-4-1989 (1989CCL-III-274).

2. Definitions - (1) In this Act, unless the context otherwise requires,-

(a) 'employer', in relation to a Scheduled Act, which defines such expression, has the same meaning assigned to it in that Act, and in relation to any other Scheduled Act, means the person who is required to furnish returns or maintain registers under that Act;

Notes

'Employer' includes a legal representative of a deceased employer- *See* Payment of Wages Act, 1936, Section 2*(i-a)*. When there is a manager who is entrusted with the affairs of the company the directors of the company cannot be said to be employer. *Superintendent and Rememberancer of Legal Affairs v. Balai Chand Saha,* 78 CWN 757: 45 FJR 489.

(b) 'establishment' has the meaning assigned to it in a Scheduled Act, and includes-

(i) an 'industrial or other establishment' as defined in Section 2 of the Payment of Wages Act, 1936 (4 of 1936).

(ii) A 'factory' as defined in Section 2 of the Factories Act, 1948 (63 of 1948).

(iii) A factory, workshop or place where employees are employed or work is given out to workers, in any scheduled employment to which the Minimum Wages Act, 1948 (11 of 1948), applies;

(iv) a 'plantation' as defined in Section 2 of the Plantations Labour Act, 1951 (69 of 1951); and

(v) a 'newspaper establishment' as defined in Section 2 of the Working Journalists and other Newspaper Employees (Conditionsof Service) and Miscellaneous Provisions Act, 1955 (45 of 1955);

S.4]
Notes

Factory- If manufacturing process is done at two different places, they are factories (AIR 1955 All 702). Premises include land as well (AIR 1956 Bom 219). Kitchen is not factory (1941) 2 KB 232, 238 (B); but a contrary view has been expressed in 1980 Lab IC 100(Bom). There is difference between the definition of factory as given in the Indian Factories Act and the English Factories Act. Whether any restaurant is a factory has to be decided after taking into account all relevant considerations, *viz.,* Section 21 *(k)* and *(m)*. Mere existence of frigidaire will not make the premises factory (AIR 1956 Mad 600). The process of grabling paper with the aid of more than twenty persons in certain premises shall make the premises a factory (58 Cr LJ 1026).

Newspaper establishment- 'Newspaper establishment' means an establishment under the control of any person or body of persons, whether incorporated or not, for the production or publication of one or more newspapers or for conducting any news agency or syndicate—See Working

Jounalists and other Newspaper Employees (Conditions of Service) and Miscellaneous Provisions Act, 1955, Section 2(d).

(c) 'Form' means a Form annexed to this Act;

(d) 'Scheduled Act' means as Act specified in the Schedule and is in force on the commencement of this Act in the territories to which such Act extends generally, and includes the rules made thereunder;

(e) 'small establishment' means an establishment in which not less than ten and not more than nineteen persons are employed or were employed on any day of the preceding twelve months;

(f) 'very small establishment' means an establishment in which not more than nine persons are employed or were employed on any day of the preceding twelve months.

3. Amendment of certain labour laws- On and from the commencement of this Act, the Scheduled Acts shall have effect subject to the provisions of this Act.

4. Exemption from returns and registers required under certain labour laws- (1) On and from the commencement of this Act it shall not be necessary for an employer in relation to any small establishment or very small establishment to which a Scheduled Act applies to furnish the returns or to maintain the registers required to be furnished or maintained under that Scheduled Act:

Provided that such employer,-

(a) furnishes, in lieu of such returns, a Core Return in Form A;

(b) maintains, in lieu of such registers,-

(i) registers in Form B, Form C and Form D, in the case of small establishments; and

(ii) registers in Form D and Form E, in the case of very small establishments;

Provided further that every such employer shall continue to-

(a) issue wage slips in the Form prescribed in the Minimum Wages (Central) Rules, 1950 made under Section 18 and 30 of the Minimum Wages act, 1948 (11 of 1948) and slips relating to measurement of the amount of work done by piece-rated workers required to be issued under the Payment of Wages (Mines) Rules, 1956 made under Section 13-A and 26 of the Payment of Wages Act, 1936 (4 of 1936); and

(b) file returns relating to accidents under Section 88 and 88-A of the Factories Act, 1948 (63 of 1948) and Sections 32-A and 32-B of the Plantations Labour Act, 1951 (69 of 1951).

(2) Save as provided in sub-section (1), all other provisions of a Scheduled Act, including in particular, the inspection of the registers by, and furnishing of their copies to, the authorities under that Act, shall apply to the returns and registers required to be furnished or maintained under this Act as they apply to the returns and registers under that Scheduled Act.

(3) Where an employer in relation to a small establishment or very small establishment to which a Scheduled Act applies, furnishes returns or maintains the registers as provided in the proviso to sub-section (1), nothing contained in that Scheduled Act shall render him liable to any penalty for his failure to furnish any return or to maintain any register under that Scheduled Act.

5. Savings- The commencement of this Act shall not affect,-

(a) the previous operation of any provision of any Scheduled Act or the validity, invalidity, effect or consequence of anything done or suffered under that provision, before the relevant period;

(b) any right, privilege, obligation or liability already acquired, accrued or incurred under any Scheduled Act, before the relevant period;

(c) any penalty, forfeiture, or punishment incurred or inflicted in respect of any offence committed under any Scheduled Act, before the relevant period;

(d) any investigation, legal proceeding or remedy in respect of any such right, privilege, obligation, liability, penalty, forfeiture or punishment aforesaid, and such investigation, legal proceeding or remedy in respect of any such right, privilege, obligation, liability, penalty, forfeiture or punishment shall be instituted, continued or disposed of, as the case may be, in accordance with that Scheduled Act.

Explanation- For the purpose of this section, the expression "relevant period" means the period during which and establishment is or was a small establishment or a very small establishment under this Act.

6. Penalty- Any employer who fails to comply with the provisions of this Act, shall, on conviction, be punishable-

(a) in the case of the first conviction, with fine which may extend to rupees five thousand; and

(b) in the case of any second or subsequent conviction, with imprisonment for a period which shall not be less than one month but which may extend to six months or with fine which shall not be less than rupees ten thousand but may extend to rupees twenty-five thousand, or with both.

7. Power to amend Form- (1) The Central Government may, if it is of opinion that it is expedient so to do, by notification in the Official Gazette amend any Form and thereupon such Form shall, subject to the provisions of sub-section (2), be deemed to have been amended accordingly.

(2) Any notification issued under sub-section(l) shall be laid before each House of Parliament, if it is sitting as soon as may be after the issue of the notification, and if it is not sitting, within seven days of its re-assembly and the Central Government shall seek the approval of Parliament to the notification by a resolution moved within a period of fifteen days beginning with the day on which the notification is so said before the House of the People, and if Parliament makes any modification in the notification or directs that the notification should cease to have effect, the notification shall thereafter have effect only in such modified form or be on no effect, as the case effect, as the case may be, but without prejudice to the validity of anything previously done thereunder.

8. Power to remove difficulties- If any difficulty arises in giving effect to the provisions of this Act, the Central Government may, by order, not inconsistent with the provisions of this Act, remove the difficulty.

Provided that no such order shall be made after the expiry of a period of two years from the date on which this Act receives the assent of the President.

The Schedule

[See Section 2(1) (d)]

(1) The Payment of Wages Act, 1936 (4 of 1936).
(2) The Weekly Holidays Act, 1942 (18 of 1942).
(3) The Minimum Wages Act, 1948 (11 of 1948).
(4) The Factories Act, 1948 (63 of 1948).
(5) The Plantations Labour Act, 1951 (69 of 1951).
(6) The Working Journalists and other Newspaper Employees (Conditions of Service) and Miscellaneous Provisions Act, 1955 (45 of 1955).
(7) The Contract Labour (Regulation and Abolition) Act, 1970 (37 of 1970).
(8) The Sales Promotion Employees (Conditions of Service) Act, 1976 (11 of 1976).
(9) The Equal Remuneration Act, 1976 (25 of 1976).

Form A

[See Section 4(1) proviso (a)]
Core Return
RETURN FOR THE YEAR ENDING 31st DECEMBER
(To be furnished on or before the 15th February of the succeeding year by small establishments and very small establishments)

1. (a) Name and postal address of the establishment.
   (b) Name and residential address of the employer.
   (c) Name and residential address of the Manager or person responsible for supervision and control of the establishment.
   (d) Name of the principal employer in the case of a contractor's establishment.
   (e) Date of commencement of the establishment.

   Nature of Operation/Industry/ Work Carried on
2. (a) Number of days worked during the year.
   (b) Number of man-days worked during the year.
   (c) Daily hours of work.
   (d) Day of weekly holiday.

3. (a) Average number of persons employed during the year
      (i) Males.
      (ii) Females.
      (iii) Adolescents (those who have not completed 14 years of age).
   (b) Maximum number of workers employed on any day during the year.
   (c) Number of worker discharged, dismissed, retrenched or whose services were terminated during the year.

4. Rates of wages- categorywise:
   (1) Males (2) Females (3) Adolescents (4) Children
5. Gross wages paid:
      (a) in cash;
      (b) in kind.
6. Deductions:
   (a)   Fines.
   (b)   Deductions for damage or loss.
   (c)   Other deductions.

7. Number of workers who were granted leave with wages during the year.

8. Nature of welfare amenities provided : Statutory (specify the statute).

9. Does the establishment carry out any hazardous process or dangerous operation coming within the meaning of the Factories Act, 1948 If so, give particulars.

10. Number of Accidents:

   (a)   Fatal.

   (b)   Non-fatal.

11. Nature of safety measures provided as required under the Factories Act, 1948.

Date.....................................*Signatue of the Employer with full name in capitals*

Place.....................

Form B

[*See* Section 4(1) proviso (*b*) (i)]

Register of Wages required to be maintained by Small Establishments
(To be maintained within seven days of the expire of the wage period)
Name of establishment..Name and address of employer...........................
Address (Local)...............................Name of work.....................................
(Permanent)..Wage period...................................

SI. No.	Name of the employee	Sex	Designation	Classification, whether permanent/ temporary/casual/part-time or any other	Father's or husband's name	Total days number of unit worked
1	2	3	4	5	6	7

WAGES EARNED

Basic wages	Dearness allowance	Overtime	Bonus or exgratio	Maternity benefits	Gratuity	Any other allowance	Total amount
Statutory minimum rate	Actual						
8	9	10	11	12	13	14	15

DEDUCTIONS

Other deductions indicating the nature	Total deductions	Net amount payable	Signature or thumb impression of employee with date	Signature of Inspector with date	Remarks
23	24	25	26	27	28

Notes : 1. In case of deduction of any advance taken by an employee, the employer shall also indicate theirin the number of installments paid/total installments by which advance is to be repaid such as "5/20, 6/20" etc. The purpose of advance shall also be mentioned in the Remarks column.

2. In case of imposition of fines or deduction for damage or loss, the specific act or omission for which the penalty has been imposed has to be indicated in the Remarks column. A certificate shall also be recorded in the said column to the effect that on opportunity to show cause was given to the employee concerned before imposition of fine or deduction.

Date....................................*Signature of the Employer with full name in capitals*
Place.................

Form C
[See Section 4(1) proviso (b)(i)
Muster-Roll to be maintained by small establishments
Name of establishment...Name and address of the employer.......................................
Address (Local)..
(Permanent).....................................Wage period.....................................

SI. No.	Name of the employee	Date of employment	Permanent address	Age or date of birth
1	2	3	4	5

Father's or husband's name	For the period ending........ Number of units of work done during	Total attendance	Total overtime worked 1	Total production in case of piece-rated workers 2
6	7	8	9	10

Compensatory rest3			
Brought forward from previous wage period	Given during the wage period	Signature of Inspector with date	Remarks
11	12	13	15

Notes : 1. In the case of daily-rated workers, the extent of overtime done on each occasion has to be reflected against each concerned date, such as "P/V meaning "Present with one hour's overtime", "P/ll/2" meaning "Present with one-and-a-half hour's overtime" and so on.

2. The number of units of work done by a piece-rated worker has to be noted for each day in the Register. In case of employment of any child/adolescent the employer shall indicate the hours worked each day with intervals of rest.

3. The compensatory rest availed by the worker has to be marked in the Register in red ink as 'CR'

4. Column 7 to be filled up on each working day and the remaining columns to be completed within seven days of the expiry of the wage period.

Date....................................*Signature of the Employer with full name in capitals*
Place....................................

Form D

[See Section 4(1) proviso (b)(i) and (ii)]
Monthly Register showing welfare amenities to be maintained by small establishments and very small establishments

SI. No.	Name of the employee	Sex	Designation	Weekly day of rest	Dates of holidays for festivals or similar other occasions
1	2	3	4	5	6

Number of casual leave availed by the employee.	Quantum of annual leave with wages		Whether welfare amenities provided for		
	Due	Availed	Rest-room	Drinking-water	First aid
7	8	9	10	11	12

Whether Scheduled Caste/ Scheduled Tribe, Handicapped, or any other particular category	Signature of the employer or his agent	Remarks of the Inspecting Officer	Signature of Inspector with date
13	14	15	16

Note:- To be completed within seven days of the expiry of each calendar month.

Date....................................*Signature of the employer with full name in capitals*

Place....................................

Form E

[See Section 4(1) proviso(b)(ii)]

Monthly register of muster roll-cum-wages required to be maintained by very small establishments

Year................................

Month................................

or Wage period................................

(where different)................................

Name of establishment................................

Name of employee................................ Father's name................................

Name of work................................Rate of wages................................

Wage period................................Date of employment................................

Date	Hours of work	Interval for Rest and Meal		Hours worked with the employer	Overtime		Casual or sickness leave availed during the month/ wage period
		From	To		Hours worked	Wages earned	
	From	To	From	To			
1	2	3	4	5	6	7	8	9

Privilege Leave					Remuneration Due			
Leave due	Leave availed	Balance	Signature of the employer	Remarks of the employer	Basic salary or wage	Over time	Other allowances, if any	Total
10	11	12	13	14	15	16	17	18

Fines and deductions on account of damage or loss by neglect or default	Deductions				Net amount of payment	Date of payment	Signature or thumb-Impression of the employee	Signature of Inspector with remarks, if any, and date
	Other deductions	Advance paid, if any						
		Date	Amount	Total				
19	20	21	22	23	24	25	26	27

Note:- Columns 1 to 12 to be filled up on each working day and the remaining columns to be completed within seven days of the expiry of the wage period.

Date.....................................*Signature of the Employee with full name in capitals*

Place...................................

Source: Tea Board of India.

Appendix Ib

The Plantations Labour (Amendment) Act, 2010

(No. 17 OF 2010)

[18th May, 2010.]

An Act further to amend the Plantations Labour Act, 1951.

BE it enacted by Parliament in the Sixty-first Year of the Republic of India as follows:—

1. (1) This Act may be called the Plantations Labour (Amendment) Act, 2010.

(2) It shall come into force on such date as the Central Government may, by notification in the Official Gazette, appoint, and different dates may be appointed for different provisions of this Act and for different States and any reference in any such provision to the commencement of this Act shall, in relation to any State, be construed as a reference to the coming into force of that provision in that State.

2. In section 2 of the Plantations Labour Act, 1951 (hereinafter referred to as the principal Act),—

   (a) in clause (e), the following *Explanation* shall be inserted, namely:—
   '*Explanation.*—For the purposes of this clause, "the person who has the ultimate control over the affairs of the plantation" means in the case of a plantation owned or controlled by—

      (i) a company, firm or other association of individuals, whether incorporated or not, every director, partner or individual;

(ii) the Central Government or State Government or any local authority, the person or persons appointed to manage the affairs of the plantation; and

(iii) a lessee, the lessee;';

(b) in clause (ee), for the words "and includes, where the worker is a male, his parents dependent upon him", the words "and includes parents and widow sister, dependent upon him or her" shall be substituted;

(c) in clause *(k)*,—

(i) in the opening portion, after the words "manual or clerical", the words "and includes a person employed on contract for more than sixty days in a year" shall be inserted;

(ii) in sub-clause (ii), for the words "rupees seven hundred and fifty", the words "rupees ten thousand" shall be substituted;

(iii) in sub-clause (iii), for the words "managerial capacity, notwithstanding that his monthly wages do not exceed rupees seven hundred and fifty", the words "managerial or administrative capacity, notwithstanding that his monthly wages do not exceed rupees ten thousand" shall be substituted.

3. In section 7 of the principal Act, in sub-section (2), in clause (b), for the words "and children are, or are to be", the word "are" shall be substituted.

4. In section 10 of the principal Act, in sub-section (2), for the words "chief inspector", the words "State Government upon a request by the chief inspector" shall be substituted.

5. After Chapter IV of the principal Act, the following Chapter shall be inserted, namely:—

CHAPTER IV A

Provisions as to safety

18A. (1) In every plantation, effective arrangements shall be made by the employer to provide for the safety of workers in connection with the use, handling, storage and transport of insecticides, chemicals and toxic substances.

(2) The State Government may make rules for prohibiting or, restricting employment of women or adolescents in using or handling hazardous chemicals.

(3) The employer shall appoint persons possessing the prescribed qualifications to supervise the use, handling, storage and transportation of insecticides, chemicals and toxic substances in his plantation.

(4) Every employer shall ensure that every worker in plantation employed for handling, mixing, blending and applying insecticides, chemicals and toxic substances, is trained about the hazards involved in different operations in which he is engaged, the various safety measures and safe work practices to be adopted in emergencies arising from spillage of such insecticides chemicals and toxic substances and such other matters as may be prescribed by the State Government.

(5) Every worker who is exposed to insecticides, chemicals and toxic substances shall be medically examined periodically, in such manner as may be prescribed, by the State Government.

(6) Every employer shall maintain health record of every worker who is exposed to insecticides, chemicals and toxic substances which are used, handled, stored or transported in a plantation, and every such worker shall have access to such record.

(7) Every employer shall provide—

    (a) washing, bathing and cloack room facilities; and
    (b) protective clothing and equipment,

to every worker engaged in handling insecticides, chemicals or toxic substances in such manner as may be prescribed by the State Government.

(8) Every employer shall display in the plantation a list of permissible concentrations of insecticides, chemicals and toxic substances in the breathing zone of the workers engaged in the handling and application of such insecticides, chemicals and toxic substances.

(9) Every employer shall exhibit such precautionary notices as may be prescribed by the State Government indicating the hazards of insecticides, chemicals and toxic substances.

18B. (1) The State Government may, by notification in the Official Gazette, make rules to carry out the purposes of this Chapter.

(2) In particular, and without prejudice to the generality of the foregoing power, such rules may provide for all or any of the following matters, namely:—

    (a) the restriction on employment of women and adolescents for handling hazardous chemicals under sub-section (2) of section 18A;
    (b) the qualifications of superviser appointed under sub-section (5) of section 18A;
    (c) the matters for training of workers under sub-section (4) of section 18A;
    (d) the medical examination of workers under sub-section (5) of section 18A;

(e) the facilities and equipment to be provided to the workers engaged in handling insecticides, chemicals and toxic substances under sub-section (7) of section 18A;

(f) the precautionary notices to be exhibited under sub-section (9) of section 18A.".

6. In section 19 of the principal Act, in sub-section (1), the words "or child" shall be omitted.

7. After section 23 of the principal Act, the following section shall be inserted, namely:—

"24. No child shall be employed to work in any plantation.".

8. In section 25 of the principal Act,—

(a) the words "or child" shall be omitted;

(b) in the marginal heading, the words "and children" shall be omitted.

9. In section 26 of the principal Act,—

(a) in the opening portion, the words "child and no" shall be omitted;

(b) in clause (b), the words "child or" shall be omitted.

10. In section 27 of the principal Act, in sub-section (1), the words "either as a child or" shall be omitted.

11. After section 32B of the principal Act, the following section shall be inserted, namely:—

"32C. The employer shall give compensation to a worker in plantation in case of accident and the memorandum relating to such compensation shall be got registered by the employer with the Commissioner in accordance with the provisions of the Workmen's Compensation Act, 1923.".

12. In sections 33, 35 and 36 of the principal Act, for the words "three months, or with fine which may extend to five hundred rupees, or with both", wherever they occur, the words "six months, or with fine which may extend to ten thousand rupees, or with both" shall be substituted.

13. In section 34 of the principal Act, for the words "one month, or with fine which may extend to fifty rupees, or with both," the words "two months, or with fine which may extend to one thousand rupees, or with both" shall be substituted.

14. In section 37 of the principal Act, for the words "six months, or with fine which may extend to one thousand rupees, or with both", the words "one year, or with fine which shall not be less than ten thousand rupees but which may extend to one lakh rupees, or with both" shall be substituted.

15. For section 39 of the principal Act, the following sections shall be substituted, namely:—

"39. No court shall take cognizance of any offence under this Act except on a complaint made by any worker or an office bearer of a trade union of which such worker is a member or an inspector and no court inferior to that of a metropolitan magistrate or a judicial magistrate of the first class shall try any offence punishable under this Act.

39A. No suit, prosecution or other legal proceeding shall lie against any person for anything which is in good faith done or intended to be done under this Act."

16. In section 43 of the principal Act, for sub-section (3), the following sub-section shall be substituted, namely:—

"(3) Every rule made by the State Government under this Act shall, as soon as may be after it is made, be laid before the State Legislature."

V.K. Bhasin,
Secy. to the Govt. of India

Source: indiacode.nic.in/fullact1.asp?tfnm=201017.

Appendix II

TABLE A1
Quantum of Tea Produced, Consumed and Exported in India: 1971–2002

(in '000 kg)

Year	Production ('000 kg)	Consumption ('000 kg)	Exports ('000 kg)	Value ('000 ₹)
1971	434,500	221,000	202,052	1,536,678
1972	454,600	233,000	198,195	1,51,1439
1973	470,800	244,000	188,192	1,427,072
1974	488,100	258,000	210,583	1,927,972
1975	485,800	272,000	218,128	2,446,592
1976	510,400	287,000	233,611	2,731,364
1977	555,400	302,000	229,637	5,416,157
1978	563,100	320,000	176,051	3,628,916
1979	543,300	332,000	199,639	3,618,401
1980	569,600	346,000	224,026	4,290,277
1981	560,000	360,000	241,246	4,342,541
1982	560,700	372,000	189,933	3,555,525
1983	581,500	386,000	208,476	5,168,145
1984	639,800	400,000	217,040	7,404,551
1985	656,200	415,000	214,021	6,952,966
1986	620,800	431,000	203,149	5,808,510
1987	665,300	446,000	201,891	6,392,466
1988	700,000	462,000	200,956	6,126,937
1989	688,100	480,000	211,622	8,400,923
1990	720,300	500,000	209,085	11,041,507
1991	754,200	524,000	201,720	11,203,136
1992	732,300	540,000	173,358	9,717,402
1993	760,800	537,000	173,726	11,325,271
1994	752,900	550,000	149,317	9,652,334
1995	756,000	562,000	167,143	11,908,077
1996	780,200	580,000	160,004	12,070,978
1997	810,000	597,000	200,713	17,218,472
1998	874,100	615,000	207,640	22,383,087
1999	825,900	633,000	189,092	19,024,369
2000	846,900	653,000	204,353	18,270,324
2001	853,900	673,000	179,857	16,022,060
2002	838,500	693,000	198,087	16,697,757

Source: Tea Statistics (Tea Board, various years).

TABLE A2
Unit Export Prices of Tea of Major Producing Exporting Countries: 1999–2007

(in US $ per kg)

Name of the Countries	1999	2000	2001	2002	2003	2004	2005	2006	2007
India	2.38	2.04	1.95	1.79	1.97	2.06	2.09	2.03	2.45
Bangladesh	1.35	1.28	1.24	1.19	1.29		1.28	1.42	1.24
Sri Lanka	2.28	2.36	2.28	2.24	2.25	2.41	2.58	2.64	3.26
Indonesia	0.99	1.06	1.00	1.03	1.09	1.18	1.19	1.41	1.51
Kenya	1.93	2.12	1.67	1.58	1.63	1.61	1.59	2.07	1.99
Malawi	0.93	0.85	0.93	0.88	0.90	1.01	1.10	1.17	1.1
Mauritius	6.02	4.37	5.27	5.72	5.63	8.29	7.57	8.44	7.41
Rwanda	1.55	1.77	1.49	1.51	1.59	1.66	1.59	1.99	0
Tanzania	1.14	1.45	1.27	1.22	1.21	1.24	1.15	1.37	1.28
Argentina	0.75	0.76	0.71	0.69	0.57	0.59	0.66	0.69	0.72
Brazil	1.83	1.75	1.58	1.40	1.35	1.52	1.71	1.78	1.94
Papua & New Guinea	1.25	1.10	1.15	1.31	1.25	1.08	1.16	1.14	1.14
Japan	14.33	15.87	13.11	14.60	16.29	18.15	18.34	16.37	16.47
Turkey	1.08	0.91	0.85	0.85	0.77	1.16	1.14	1.15	1.2
Uganda	1.36	1.38	1.09	1.12	1.00	1.56	1.00	1.10	1.35
Zimbabwe	1.31	1.09	0.38						
China	1.7	1.52	1.37	1.21	1.38	1.56	2.06	1.88	2.10

Source: Annual Bulletin of Statistics, 2008, International Tea Committee, London.

TABLE A3
Trend in Auction Prices in Major Tea Producing Countries

(US $ per kg)

Year	Indian Tea at Indian Auction	Sri Lankan Tea at Colombo Auction	African Tea at Mombassa Auction
2000	1.37	1.75	2.02
2003	1.20	1.54	1.54
2004	1.42	1.78	1.55
2005	1.32	1.84	1.47
2006	1.46	1.90	1.93
2007	1.62	2.51	1.66
2008 (P)	1.99	2.83	2.18

Source: Tea Board of India.
Note: (P) Provisional

TABLE A4
World Demand and Supply of Tea

(Figures in million kg)

Year	World Supply	World Demand	(+) or (−)
2004	3334.53	3192.93	(+) 141.60
2005	3457.59	3348.52	(+) 109.07
2006	3572.66	3466.99	(+) 105.67
2007	3802.94	3710.84	(+) 92.10
2008	3749.78	3596.17	(+) 153.61

Source: Tea Board of India.

Bibliography

Ahluwalia, Montek S. 2000. 'Economic Performance of States in Post-reforms Period', *Economic and Political Weekly*, 35(19): 1637–48.

———. 2002. 'State Level Performance under Economic Reforms in India', in Anne Krueger (ed.), *Economic Policy Reforms and the Indian Economy*. Chicago: University of Chicago Press.

Anant, T.C.A. 2009. 'Revisiting Labour Market Regulation', *Indian Journal of Labour Economics*, 52(2): 195–202.

Anant, T.C.A., R. Hassan, P. Mohapatra, R. Nagaraj, and S.K. Sasikumar. 2006. *Labour Markets in India: Issues and Perspectives in Labour Markets in Asia*. London: Macmillan.

Baak, P.E. 1991. 'The Conference on Capitalist Plantations in Colonial Asia', *South Asia Research*, 11(1): 89–95.

Banerjee, K. et al. 2003. The Report of the Fact Finding Team That Visited Tea Plantations in Kerala and Tamil Nadu, Centre for Education and Communication, New Delhi.

Beck, U. 2000. *What is Globalization?* Cambridge: Polity Press.

Behal, Rana P. 2003. *Wage Structure and Labour: Assam Valley Tea Plantations, 1900–1947*, VVGNLI Research Study Series, 043/2002, VV Giri National Labour Institute, Noida.

———. 2006. 'Power Structure, Discipline, and Labour in Assam Tea Plantations under Colonial Rule', *IRSH* 51, Supplement, pp. 143–72.

Behal, Rana P. and Prabhu Mohapatra. 1992. '"Tea and Money versus Human Life": The Rise and Fall of the Indenture system in the Assam Tea Plantations 1840–1908', *The Journal of Peasant Studies*, 19(3–4): 142–71.

Besky, Sarah. 2008. 'Can a Plantation be Fair? Paradoxes and Possibilities in Fair Trade Darjeeling Tea Certification', *Anthropology of Work Review*, 29(1): 1–9.

Beteille, Andre. 1981. 'Foreword' in Sharit Bhowmik, *Class Formation in the Plantation System*, Delhi: People's Publishing House.

Bhattacharya B.B. and S. Sakthivel. 2004. 'Regional Growth and Disparity in India: Comparison of Pre- and Post-Reform Decades', *Economic and Political Weekly,* 39(10): 1071–77

Bhowmik, S.K. 1981. *Class Formation in the Plantation System*, People's Publishing House, Delhi.

Bhowmik, S.K. 2002. 'Productivity and Labour Standards in Tea Plantation Sector in India', in A. Sivananthiran and C.S. Venkata Ratnam (ed.), *Labour and Social Issues in Plantations in South Asia: Role of Social Dialogue*. New Delhi: ILO – SAAT and IIRA..

Biswas, S., Debasish Chokraborty, Sutay Berman, R.N., and Joshua Berman. 2005. 'Nutritional Survey of Tea Workers on Closed, Reopened, and Open Tea Plantations of the Dooars Region, West Bengal, India', West Bengal Agricultural Workers' Union with International Union of Foodworkers and American Jewish World Service, Jalpaigudi.

Bose, S. 1954. *Capital and Labour in the Indian Tea Industry.* Bombay: AITUC Publications.

Brass, Tom and Henry Bernstein. 1992. 'Introduction: Proletarianisation and Deproletarianisation on the Colonial Plantation', *The Journal of Peasant Studies,* 19(3–4): 1–40.

Das Gupta, Ranjit. 1992a. 'From Peasants and Tribesmen to Plantation Workers: Colonial Captialism, Reproduction of Labour Power and Proletarianisation in North East India, 1850s to 1947', *Economic and Political Weekly,* 21(4): PE2–PE10.

———. 1992b. 'Plantation Labour in Colonial India', *The Journal of Peasant Studies,* 19(3–4): 173–97.

———. 1992c. 'Plantation Labour in Colonial India' in E. Valentine Daniel, Henry Bernstein and Tom Brass (eds), *Plantations, Peasants and Proletarians in Colonial Asia.* London: Frank Cass.

Das, K. 2002. *Labour Contracts and Work Agreements in Tea Plantations of Assam,* NLI Working Paper Series No. 033, V. V. Giri National Labour Institute, Noida, India.

Dasgupta, Keya. 1983. 'Plantation Economy and Land Tenure System in Brahamaputra Valley: 1893–1914', *Economic and Political Weekly,* 18(29): 1280–90.

Davis, Junior R. 2003. 'The Rural Non-farm Economy, Livelihoods and their Diversification: Issues and Options', NRI Report No. 2753, Natural Resource Institute, Kent.

De Haan, Leo and Zoomers, Annelies. 2005. 'Exploring the Frontier of Livelihoods Research', *Development and Change,* 36(1): 27–47.

De Jocas, Yves and Guy Rocher. 1957. 'Inter-Generation Occupational Mobility in the Province of Quebec', *The Canadian Journal of Economics and Political Science,* 23(1): 57–68.

Deshpande, L.K., A.N. Sharma, Anup K. Karan and Sandip Sarkar. 2004. *Liberalisation and Labour: Labour Flexibility in Indian Manufacturing.* New Delhi: Institute of Human Development.

Dyer, Graham. 1996. 'Output Per Acre and Size of Holding: The Logic of Peasant Agriculture under Semi-Feudalism', *Journal of Peasant Studies,* 24(1–2): 103–31.

Ellis, F. 1998. 'Households Strategies and Rural Livelihood Diversification', *Journal of Development Studies,* 35(1): 1–35.

———. 1999. 'Rural Livelihood Diversity in Developing Countries: Evidence and Policy Implications', *Natural Resource Perspectives,* 40.

———. 2000. 'The Determinants of Rural Livelihood Diversification in Developing Countries', *Journal of Agricultural Economics*, 51(2): 289–302.

Evans, Barbara. 1995. 'Constructing A Plantation Labour Force: The Plantation-Village Nexus in South India', *The Indian Economic and Social History Review*, 32(2): 155–76.

Fafchamps, M. and B. Minten. 1999. 'Social Capital and the Firm: Evidence from Agricultural Trade', www.economics.ox.ac.uk/members/marcel.fafchamps/homepage/social.pdf

Fay, C.R. 1936. 'Plantation Economy', *The Economic Journal*, 46(184): 620–44.

Fernades, W., S. Barbora and G. Bharali. 2003. *Children of the Plantation Labourers and their Right to Education*. Guwahati: North Eastern Social Research Centre.

Ferreira, F.G.H. and Peter Lanjouw. 2001. 'Rural Non-farm Activities and Poverty in the Brazillian North East', *World Development*, 29(3): 502–28.

Genovese, E.D. 1967. *The Political Economy of Slavery: Studies in the Economy and Society of the Slave South*. New York: Vintage.

Ghosh, Kaushik. 1999. 'A Market for Aboriginality: Primitivism and Race Classification in the Indentured Labour Market of Colonial India', in Gautam Bhadra, Gyan Prakash and Susie Tharu (eds), *Subaltern Studies X. Writings on South Asian History and Society*. New Delhi: Oxford University Press.

Ginsberg, Morris. 1929. 'Interchange between Social Classes', *The Economic Journal*, 39(156): 554–65.

Glyn, Andrew. 2006. *Capitalism Unleashed: Finance, Globalisation and Welfare*. Oxford: Oxford University Press.

Gohain, H. 2007. 'A Question of Identity: Adivasi Militancy in Assam', *Economic and Political Weekly*, 42(49): 13–16.

GoI. 2007. *Report of Committee on Legislation Plantation Sector*. New Delhi: Department of Commerce, Ministry of Commerce & Industry, Government of India.

Graham, E. and I. Floering. 1984. *The Modern Plantation in the Third World*. Kent: Croom Helm.

Gregor, Howard F. 1965. 'The Changing Plantation', *Annals of the Association of American Geographers*, 55(2): 221–38.

Guha, Amalendu. 2006. *Planter Raj To Swaraj: Freedom Struggle And Electoral Politics In Assam 1826–1947*. Delhi: Tulika Books.

Harriss-White, Barbara. 2002. 'Globalisation, Insecurities and Responses: An Introductory Essay' in Barbara Harriss-White (ed.), *Globalization and Insecurity: Political, Economic and Physical Challenges*. London: Palgrave Macmillan.

———. 2003. *India Working*. Cambridge: Cambridge University Press.

———. 2009. 'Agro-Capital and the Local Foodgrains Economy in Northern Tamil Nadu, 1973–2009', paper presented at the international workshop on 'Market Town, Market Society, Informal Economy, Oxford University, 8–9 June 2009 (unpublished).

Harriss-White, Barbara and Gooptu, Nandini. 2000. 'Mapping India's World of Unorganized Labour', in Leo Panitch and Colin Leys (eds), *Working Classes Global Realities: Socialist Register 2001*. London: The Merlin Press.

Hauser, Robert M. 1980. 'Some Exploratory Methods for Modeling Mobility Tables and Other Cross-Classified Data', *Sociological Methodology*, 11(1): 413–58.

Held, David, A. Mcgrew, D. Goldblatt and Jonathan Perraton. 1999. *Global Transformations: Politics, Economics and Culture*. Cambridge: Polity Press.

Held, David and A. Mcgrew, eds. 2003. *The Global Transformations Reader: An Introduction to the Globalization Debate*. Cambridge: Polity Press.

Helleiner, Gerald K. 2000. 'Markets, Politics and Globalization: Can the Global Economy be Civilized?' The Tenth Raúl Prebisch Lecture, Geneva, 11 December 2000, CIS Working Paper 2000-1, http://www.uoit.ca/sas/Globalization%20and%20WTO/MarkPolGlob.pdf

Hensman, Rohini. 2001. The Impact of Globalisation on Employment in India and Responses from the Formal and Informal Sectors, CLARA Working Paper No.15, http://www.iisg.nl/clara/publicat/clara15.pdf. Last accessed 29 January 2011.

Hirst, P. 1997. 'The Global Economy: Myths and Realities', *International Affairs*, 73(3): 409–25.

Hirst, P. and G. Thompson. 1996. *Globalization in Question: The International Economy and the Possibilities of Governance*. Cambridge: Polity Press.

Huws, Ursula. 2006. 'Fixed, Footloose, or Fractured: Work, Identity, and the Spatial Division of Labor in the Twenty-First Century', *Monthly Review*, 57(1). Available from: http://monthlyreview.org/2006/03/01/fixed-footloose-or-fractured-work-identity-and-the-spatial-division-of-labor-in-the-twenty-first-century-city. Last accessed on 29 January 2011.

Indian Tea Association (ITA). 1933. *Detailed Report of the General Committee*, Calcutta.

International Monetary Fund (IMF). 2008. 'Globalization: A Brief Overview', *Issues Brief 02/2008*, http://www.imf.org/external/np/exr/ib/2008/053008.htm. Last accessed 26 January 2011.

Jain, Shobhita. 1998. 'Gender Relations and Plantation System in Assam, India', in Shobhita Jain and R. Reddock (eds), *Women Plantation Workers: International Experiences*. Oxford and New York: Berg.

Jolliffe, D. 1999. 'The Impact of Education in Rural Ghana: Examining Productivity and Labor Allocation Effects', mimeo, Centre for Economic Research and Graduate Education, Charles University, Prague.

Karmakar, K.G. and G.D. Banerjee. 2005. *The Tea Industry in India: A Survey*, Occasional Paper 39, Department of Economic Analysis and Research, National Bank for Agriculture and Rural Development, Mumbai.

Khawas, Vimal. 2006. *Socio-Economic Conditions of Tea Garden Labourers in Darjeeling Hills.* New Delhi: Council for Social Development.

Krishnakutty, M., et al. 1993. 'Relative Share of Wages', *Economic Times*, 19 April, Mumbai.

Kumar, Pramod, P.S. Badal, N.P. Singh, and R.P. Singh. 2008. 'Tea Industry in India: Problems and Prospects', *Indian Journal of Agricultural Economics*, 63(1): 84–96.

Lahiri, S. 2000. 'Bonded Labour and the Tea Plantation Economy', *Revolutionary Democracy*, 6(2), http://www.revolutionarydemocracy.org/rdv6n2/tea.htm. Last accessed 24 January 2011.

Lanjouw, Peter. 1999. 'The Rural Non-farm Sector: A Note on Policy Options', Development Economics Research Group, The World Bank.

Leys C. 1994. *African Capitalists and Development.* Boulder, Colorado.

Lines, Thomas. 2006. 'Sustainable Livelihoods for Indian Tea Workers: The International Dimension', Tradecraft Exchange.

Long, Jason and Joseph Ferrie. 2005. 'A Tale of Two Labor Markets: Inter-generational Occupational Mobility in Britain and the U.S. Since 1850', Working Paper 11253 National Bureau of Economic Research, Cambridge http://www.nber.org/papers/w11253.

Lysandrou, P. 2005. 'Globalisation as Commodification', *Cambridge Journal of Economics*, 29(5): 769–97.

Magdoff, Harry. 2002. *Essays on Imperialism and Globalization*, New York: Monthly Review Press, Kharagpur: Cornerstone Publications.

Matras, J. 1960. 'Comparison of Intergenerational Occupational Mobility Patterns: An Application of the Formal Theory of Social Mobility', *Population Studies*, 14(2): 163–69.

Mazumdar, Dipak and S. Sarkar. 2008. *Globalization, Labour Markets and Inequality in India.* Oxon: Routledge.

Miller, P.W. and P.A. Volker. 1985. 'On the Determination of Occupational Attainment and Mobility', *Journal of Human Resources*, 20(2): 197–213.

Mishra, Deepak K. 2007. 'Gender, Forests and Livelihoods: A Note on the Political Economy of Transition in North-East India', *Social Change*, 37(4): 65–90.

Mishra, Deepak K. Vandana Upadhyay and A. Sarma. 2008. '"Crisis" in the Tea Sector: A Study of Assam Tea Gardens', *The Indian Economic Journal*, 56(3): 39–56.

Mishra, Deepak K. A. Sarma and Vandana Upadhyay. 2011. 'Invisible Chains? Crisis in the Tea industry and the 'Unfreedom' of Labour in Assam's Tea Plantations', *Contemporary South Asia*, 19(1): 75–90.

Mishra, K. 2005. 'Growing Discontent of Adivasis in Assam', *Counter Currents*, http://www.countercurrents.org/adivasi-mishra120405.htm. Last accessed on 14 May 2006.

Misra, Sibranjan. 1986. *The Tea Industry in India.* Delhi: South Asia Books.

Misra, U. 2003. 'Assam Tea: The Bitter Brew', *Economic and Political Weekly*, 38(22): 3029–32.

————. 2007. 'Adivasi Struggle in Assam', *Economic and Political Weekly*, 42(51): 11–14.

Mitra, N. 1991. 'Indian Tea Industry: Problems and Policies', *Economic and Political Weekly*, 26(48): M153–6.

Murray, Colin. 2001. *Livelihoods Research: Some Conceptual and Methodological Issues,* Background Paper 5, Chronic Poverty Research Centre, University of Manchester.

Nagraj, K. and L. Vedavalli. 2004. *Contractual Arrangements in the Tea Plantations of Tamil Nadu*, NLI Research Studies No. 054/2004, V.V. Giri National Labour Institute, Noida.

National Commission for Enterprises in the Unorganised Sector (NCEUS). 2007. *Report on Conditions of Work and Livelihoods in the Unorganised Sector.* New Delhi: Government of India.

Nayyar, Deepak. 2003. 'Work, Livelihoods and Rights', *Indian Journal of Labour Economics,* 46(1): 3–13.

Nayyar, Gaurav. 2005. 'Growth and Poverty in Rural India: An Analysis of Inter-State Differences', *Economic and Political Weekly,* 40(16): 1631–39

Ramadurai, N. 2002. 'FDI in Tea Sector', *The Hindu*, 12 August.

Raman, Ravi. 2010. *Global Capital and Peripheral Labour: The History and Political Economy of Plantation Workers in India.* London and New York: Routledge.

Rao, K.B.K. and A. Hone. 1974. 'India and the World Tea Economy', *Economic and Political Weekly*, 9(28): 1111–8.

Reddock, R. and Shobhita Jain. 1998. 'Plantation Women: An Introduction' in Shobhita Jain and R. Reddock (eds), *Women Plantation Workers: International Experiences*, Oxford and New York: Berg.

Saunders C.T. 1931. 'A Study of Occupational Mobility', *The Economic Journal*, 41(162): 227–40

Savur, M. 1973. 'Labour and Productivity in the Tea Industry', *Economic and Political Weekly,* 8(11): 551–59.

Scholte, Jan Art. 1997. 'Global capitalism and the State', *International Affairs*, 73(3): 427–52

Siddique, M.A.B. 1995. 'The Labour Market and the Growth of the Tea Industry in India: 1840–1900', *South Asia*, 18(1): 83–113.

Siddique, Md. Abu B. 1990. *Evolution of Land Grants and Labour Policies of the Government: The Growth of Tea Industry in Assam 1834–1940.* Delhi: South Asian Publishers.

Singh, Nirvikar, Laveesh Bhandari, Aoyu Chen and Aarti Khare. 2003. 'Regional Inequality in India: A Fresh Look', *Economic and Political Weekly*, 38(11): 1069–73.

Sivaram, B. 2000. 'Productivity Improvement and Labour Relations in the Tea Industry in South Asia', Working Paper, Sectoral Activities Programme, ILO, Geneva.

Skonieczny, Amy. 2010. 'Interrupting Inevitability: Globalization and Resistance', *Alternatives*, 35: 1–2.

Sweezy, Paul. 1997. 'More (or less) on Globalization', *Monthly Review,* Vol. 49, No.4, pp. 1–2.

Talwar, Anuradha, Debasish Chakraborty and Sarmishtha Biswas. 2005. *Study on Closed and Re-opened Tea Gardens In North Bengal,* Paschim Banga Khet Majoor Samity and International Union Of Food, Agriculture, Hotel, Restaurant, Catering, Tobacco, Plantation and Allied Workers' Associations (IUF).

Tea Board. 2003. *Tea Statistics,* www.teaindia.com. Last accessed 6 November 2011.

———. 2004. *Tea Statistics,* www.teaindia.com. Last accessed 6 November 2011.

The Guardian. 2007. 'Hundreds of Workers Dies as India's Tea Industry Suffers Crisis', 11 June 2007, http://www.guardian.co.uk/world/2007/jun/11/india.randeepramesh. Last accessed 16 December 2011.

The Hindu. 2010. 'Changing Contours of Indian Tea Industry', *The Hindu,* 29 November, Delhi.

The Plantations Labour (Amendment) Act. 18 May 2010, indiacode.nic.in/fullactI.asp?tfnm=201017. Last accessed 19 December 2011.

Tiffen, M. and M. Mortimore. 1990. *Theory and Practice in Plantation Agriculture: An Economic Overview.* London: Overseas Development Institute.

Upadhyay, Vandana and Deepak K. Mishra. 2004. 'Micro-enterprises in Hill Economies: The Case of Arunachal Pradesh', *The Indian Journal of Labour Economics,* 47(4): 1027–38.

Vandekerckhove, Nel. 2009. '"We are Sons of This Soil": The Endless Battle over Indigenous Homelands in Assam, India', *Critical Asian Studies,* 41(4): 523–48.

von Braun, J. and E. Diaz-Bonilla. 2008. 'Globalization of Agriculture and Food: Causes, Consequences and Policy Implications', in J. Von Braun, and E. Diaz-Bonilla (eds), *Globalization of Food and Agriculture and the Poor.* Delhi: Oxford University Press.

About the Authors

Deepak K. Mishra is Associate Professor of Economics, Centre for the Study of Regional Development, Jawaharlal Nehru University, New Delhi. He holds a Ph.D. from Jawaharlal Nehru University. Dr Mishra's interests are in the areas of agrarian change, migration and livelihood diversification in mountain economies.

Vandana Upadhyay is Assistant Professor, Department of Economics, Rajiv Gandhi University, Itanagar, Arunachal Pradesh. Her research interests include labour and employment, human development, gender, and livelihood issues in northeast India.

Atul Sarma is Visiting Professor, Institute of Human Development, New Delhi. He was formerly Professor of Economics at the Indian Statistical Institute, Delhi. Until recently, he was a Member of the 13th Finance Commission, Government of India. He has also served as Vice-Chancellor, Rajiv Gandhi University, Itanagar, Arunachal Pradesh.

Index

globalisation, 181; manifestations, 5;
    meaning of, 5
gram sabhas, 17
Guha, Amalendu, 25, 80, 98, 104–5

Harriss-White, Barbara, 9–10, 13,
    160

immigrant labour, 80
indentured labour, 12, 26, 104, 185,
    188
Indian Tea Association (ITA), 99
Industrial Disputes Act, 1947, 190
Industrial employment (Standing
    Orders) Act, 1946, 190
industrial plantation system, 161
informal sector: employment in, 1
Information and Communications
    Technologies (ICTs), 7
in kind wages, 108
Inland Immigration Act, 1863, 80
Inland Immigration Act of 1882,
    104
intergenerational mobility matrix:
    earlier generation: for all workers,
    in inside tea gardens, 140; for all
    workers, in outside tea gardens,
    141; inside tea gardens: all workers,
    129–30; male workers, 131–32;
    outside tea gardens: all workers,
    134–35; male workers, 136–37
intergenerational occupational
    mobility: analysed on transition
    matrices basis, 125; current
    workers, of, 127–38; determinants
    of, 142–44; earlier generations, of,
    138–42; elements: definition of,
    125; extent and pattern of: occu-
    pational distribution of fathers,
    122–24; occupational distribution
    of grandfathers, 124–25; occu-
    pational distribution of workers,
    115–22; study of, 113

International Labour Committee of
    Work on Plantations, 22–23
International Labour Organization
    (ILO), 24
International Monetary Fund (IMF),
    8

Kerala tea gardens, labour per hectare,
    87

labour: and globalisation, 5–9; glob-
    alising India, in, 9–14
labour absorption, in spatio-temporal
    variations: labour per hectare,
    analysis of, 87
labour employed, tea producing states
    and districts in, 84–85
labour households, demographic
    characteristics of, 170–71
labour, in tea gardens: employment
    of Chinese slaves, 80; trends in
    labour-use and employment:
    direct employment, to labourers,
    81; growth of employment,
    86; labour employed, 84–85;
    production, 84–85; production
    per labour, 84–85; residents
    labourers employment, 82–83
Labour Investigation Committee,
    190
labour market, in Indian economy,
    10
labour market scenario, tea gardens
    in, 185
labour per hectare: Assam district,
    in, 90; growth rate of, 91; trends
    in, 89
labour productivity, growth in,
    88–91
labour-use, trends in, 81–87
Lahiri, S., 15, 101–2
land alienation, 17
liberalisation, 181

rates of area, production and yield of, 32

tea bushes, age distribution of: area under different age group: in 1980, 72; in 1985, 73; in 1990, 73–74; in 1995, 74; in 2000, 75; in 2002, 75–76

Tea Districts Emigrant Labour Act, 1932, 190

tea estates, 16, 27

tea garden(s), 15; growth trends of, 18; size-class-wise distribution of, 52–55; size-productivity relationship in, 55–64; workers, 2; estimated daily wage rates of, 109; intergenerational occupational mobility among, 18

tea industry, in India: crisis in 1990s, 1, 14, 77–78, 182; employment elasticity, 91–93; global perspective, 28–43; growth of states, 43–48; largest producer, in world, 3; origin of, 3; position in economy, 1

tea labour force, changing composition of: in 2004, 96–97; Assam, in, 96; employment elasticity, with respect to production, 94–95

tea plantations: history of, 24–26; wage structure in: North India, 106; South India, 107

tea production, in India: growth of, 26–28

tea tribes, 17

Tiffen, M., 23–24

trade liberalisation, 13

Transport of Native Labourers Act of 1863, 104

Unfreedom, 2, 161

vertical mobility, 133

wage determination, 104–10

Wages Act, 1936, 190

wage structure, in tea plantations: North India, 106; South India, 107

West Bengal tea gardens: labour per hectare, 87

workers, desire to change: employer: outside tea garden, 153; tea garden, 152; occupation: outside tea garden, 155; tea garden, 154

workers distribution: occupation according to father occupation, in: outside tea gardens, 139; within tea gardens, 138

workers satisfaction: current occupation, in, 148; perceptions regarding: outside tea garden, 149; tea garden jobs, at outside, 151

working conditions and occupational change: perceptions on, 147–53

Workmen's Breach of Contract Act, 1859, 80

Workmen's Compensation Act, 1923, 190

World Bank, 8

world capitalism: plantation sector and (*see* Plantation sector and world capitalism)

world demand, of tea, 234

WTO Agreement, 14

For Product Safety Concerns and Information please contact our EU
representative GPSR@taylorandfrancis.com
Taylor & Francis Verlag GmbH, Kaufingerstraße 24, 80331 München, Germany